Death Rides The Sky

Incredible Survivor Stories of America's Worst Tornado

ANGELA MASON

Published by
IllinoisHistory.com
PO Box 1142
Marion IL 62959

Front Cover
West Frankfort, Ill.

Back Cover
Angela Mason and her granddaughter Number 1, Evelynn.
Portrait coursey of *Tara Eveland Photography*

Cover Design
Jon Musgrave

Library of Congress Control Number:

International Standard Book Number (ISBN)
paperback: 978-0-9891781-5-0

Printed in the United States of America
2nd Edition

Table of Contents

Laughter from the Ruins

The only tornado joke to have come out of all the research

"There was an old farmer and his old wife who had survived the great tornado of 1925, and some of their animals had come through surviving, too. One of these was their old rooster.

"The only problem was, the old rooster had lost all his feathers in the big blow. He was runnin' around the chicken yard, nekkid as he could be, and boy, was he miserable! That rooster used to really be a goer with the hens, and now, the ones that were left, even the ones that had lost their own feathers in the big blow, wouldn't give him the time of day.

"So the old farmer's wife took pity on him and decided she was going to make him his own pair of bib overalls. She sewed and she sewed and made 'em just the right size for the old rooster, and finally, she gave 'em to him and he put 'em on.

"Boy! That rooster strutted and crowed and bobbed up and down the henyard. He just looked so fine in those new bibs, and he knew it. He was a goer again, even without feathers!

"Well, he worked his way all through the old farmer's henyard, and, since there weren't that many chickens left after the big storm, he took off down the road to the next farm, where there was a young farmer and his young wife. And the old rooster just went to town on those hens there, much to their delight, 'cause they'd lost their rooster to the big blow and were in need of male companionship.

"After a few days of this, the old farmer's wife was paid a visit by the young farmer's wife.

"The young girl was just a-splittin' her sides laughing and giggling. She was laughing so hard tears were running down her face, and she could hardly speak to the old farmer's wife.

"Finally, she was able to say, 'You know, that rooster in those bib overalls, that's got to be the funniest thing I'll ever see!!'

"The old farmer's wife looked at her laughing, and said, very seriously, 'No, that's not the funniest thing you'll ever see.'

"'It's not?!?' she cried, puzzled. She still couldn't stop laughing. So she asked the old farmer's wife, 'Well, then, what could be more funny than a rooster in a pair of bib overalls?'

"'The rooster *in* his bib overalls isn't the funniest thing you'll ever see,' said the farmer's wife. 'The funniest thing you'll ever see is the rooster holding down one of the chickens with one claw, and trying to get *out* of the bibs with his other claw.

"'*That's* the funniest thing you'll ever see.'"

— courtesy of Adrian Dillon, 89, in March 2000,
Waltonville, Illinois, former resident of Parrish, Illinois
and survivor of the Tri-State Tornado of 1925

Introduction

Did you know that the wind doesn't blow the feathers off chickens in a storm?

I didn't, either, and I was raised on a farm all my early years and right through my teens, helping tend to daily activities that included "chores," where chickens were just a way of life.

I believed the "old wives' tale," if you will, about the wind plucking the chickens clean in a big blow, as that was never disputed in my family of farmers that stretched back generations on the same southeastern Illinois soil as those before them.

However, during the course of research for this book, I learned that it's not that chicken feathers can't withstand 200 mph winds while the little chicken body somehow does, claws dug into the ground and hanging on for dear life while their feathers abandon them like, well….a chicken.

Instead, chickens (hens *and* roosters; there's a difference, you know) have what could be considered a fowl's version of the human "fight or flight" reflex. In chicken terms, it could be known as "flee or be food."

Anyone who's ever plucked a chicken (and that's not a whole lot of people these days, I realize) knows that the pores (follicles) out of which the pointy-end of the feather (the quill) are pulled kind of pucker once that bird is plucked….it's that "gooseflesh" appearance. Those follicles *hold* the quill of the feather in place. When the quill is no longer there, the follicle kind of 'sits up,' and looks knobby. There's a reason for that.

It's because that flesh is literally "holding" the feather in its place. And if a chicken gets scared—such as if a predatory critter like a raccoon or fox or even another bird, like a hawk—is after it, that flesh has a reaction. The reaction is to loosen up and let go of the feathers at the quill. That way, if a raccoon is able to get close enough to close his mouth over that hen's tail feathers, he's not going to be able to yank her back—he's going to get a mouth full of feathers, because her follicles simply "let go" of the quill, and that hen is off and running, flapping and squawking and alerting the other chickens that there's a predator in their midst.

The same thing, animal science has concluded, happens in a big blow. It's not just the wind that scares the fowl at times; thunder, lightning and hail, the frequent accompaniment of tornadic storms, also causes the chickens to believe they're in mortal peril. So when a massive storm blows over, and there are chickens in the barnyard or their own enclosure (like we always had), sometimes if that storm is bad enough, the chickens will 'let go' of their feathers. The wind picks up the feathers and carries them away, leaving a huddling chicken with no feathers at all to be seen in the vicinity, and the result can be that the surviving chickens are naked...not the way God intended for them to be, and the feathers grow back, but it's a neat little trick designed for chicken self-preservation, and has grown to be one of those myths that surround violent storms sweeping across farms in the nation's Heartland.

You'd think that this would be something I would have learned, having spent my first two decades of life on a farm, but that just was one of those factoids that escaped me.

There were some other, bigger, facts and factoids that I didn't learn about until later, as well.

In the summer of 1981, I was a couple of months short of 19, and sitting in the dining room of my grandparents' house in rural Mt. Erie, Illinois, discussing with them how miserably hot the past few summers had become, and how, in my limited experience and memory, I couldn't understand why summers couldn't be like they had been even just ten years before. And why, I posed the question, had the spring and summer storms gotten so severe lately? Didn't weather used to be calmer?

My grandpa Ross Mason, 69 at the time and one of the finest men I'd ever known or have known since, was rolling a homemade cigarette and the scent of Velvet Tobacco wafted up from the can just before he sealed the lid on it.

"It's been pretty bad lately," Grandpa said and licked the rolling paper so it would seal the tobacco in. This was years before he was off the smoking habit per doctor's orders, when he eventually got down to one lung with partial function and only the top lobe of the other. He lit the home-rolled cancer stick, and the tobacco, which always smelled so delicious from the canister but not so great when burning, flamed briefly. "It's been so bad lately that it reminds me of 1925," Grandpa stated simply.

Grandma Mildred Mason, who had been removing dishes from the table after lunch, stopped in mid-reach and looked at Grandpa with an expression that could only be termed as instantly worried. My eyes narrowed as I gazed first at one grandparent, then the other. Grandma continued on with clean-up as though nothing happened, but there was suddenly something thick in the air…not quite tension, but darned close to it…more like heavy memories of something spoken of frequently, long ago, in hushed and somewhat frightened tones.

Chickens clucked loudly outside from the nearby henhouse and pigs noisily rooted their feeders.

"What happened in 1925?" I finally asked, since I could tell Grandpa wasn't going to offer up any more of his own volition.

"Big storm," Grandpa responded in a somewhat flat tone.

Grandma was noticeably absent doing dishes in the kitchen. I couldn't turn to her for assistance. Keeping with the two-word question-and-answer Grandpa had established, I continued.

"How big?" I asked.

"Killed about 700 people," Grandpa broke from the established pattern ominously.

Apparently this was not something Grandpa was used to talking about, as he was clearing his throat nervously. But he could tell he had piqued my curiosity, and once piqued, that was an intangible that took on a life of its own—a monstrous life.

But not so monstrous, I discovered, as the storm about which my Grandpa was to relate to me on that summer Sunday afternoon.

The questions I have asked myself in the years following, and what drove me to produce this material you are about to read, is, naturally, 'Why, in the age of burgeoning information and knowledge, did it take nineteen years for me to discover this storm? Where were the accounts, the annual memorials, the notoriety such an event could have and should have brought to the very area where I lived? (which was exactly one county north of some of the worst damage incurred by the storm).

'Why, in nineteen years of living, had I not heard the term *Tri-State Tornado*?'

What Grandpa spent the next hour and a half relating to me stuck with me until this day and will no doubt last into the decades to come.

He told me how his family, living north of Ellery and east of Massilon, a rural area of Edwards County just across the river from Wayne, had watched nervously the black-and-blue storm system barreling across the horizon to the south, moving southwest to northeast and directly away from them, to their great and inexpressible relief.

He told me about how several area people, my great-grandpa Willie Borah (Grandpa's future father-in-law) included, took off for the area of devastation out of curiosity but returned to send word that hundreds were in need of help, a call to which many in Wayne and Edwards counties responded.

He told me how a few days later he and one of his brothers, both of them in their early teens, had been wandering around the family property near the Little Wabash River and had come upon an entire bolt of red calico fabric. The bolt was on a roll as was the standard back then, and the fabric was in pristine condition—there was not a stain or a tear on it, and the pins that held it in place had not been disturbed. The boys took the fabric from where they had found it in the field and brought it to their mom, who, as was the habit with people back then, attempted to discover the owner of the fabric instead of just assuming it was now hers. Weeks went by before they came to the conclusion that not only was the owner probably not local, but that owner might not even be alive any longer. The bolt had likely been driven into Edwards County by prevailing winds that had kept it aloft in the hours following the storm, and had been deposited gently in the prairie grass that edged the riverbanks; there was no way of telling from whence it had originated.

Great-grandma Mason allegedly made dresses out of the fabric after all was said and done.

Throughout the month of March, 1999, the 74th anniversary of the devastating storm, working for a local newspaper, my journalistic curiosity was piqued. I formally interviewed a handful of survivors for a well-received anniversary article at that newspaper, and informally interviewed many survivors of the storm, all from within about an hour's radius of my home. I found several survivors even in my hometown. I compiled so much information and spoke so frequently of it to folks that I was repeatedly invited to speak at civic club meetings as a guest, conveying the important message of storm safety and preparedness. Somewhere along the line, someone suggested

I write a book about what I had discovered. I said "Nope. I have three kids and a demanding job; somebody else can do it."

During the Doppler Hearing in Evansville, Indiana on June 15, 1999 during which a panel of 'experts' and politicians was receiving input on whether or not to place Doppler Radar in the location of Owensville, Indiana, I had not planned to speak but had brought along the information I had compiled over the last few months, as well as one tattered copy of the well-received news article, to present to the panel (by my recollection, we had exactly NINETEEN of those issues in the archives, of 13,000 printed. It was very, very popular). Sitting through hours of horror stories of little F0 to F2 twisters that had struck the Tri-State in the years following the removal of Doppler from Evansville, I felt compelled to at last speak. One of the last to come to the microphone, I held up the tattered copy of the paper and told the panel that they were sitting just south and east of the worst tornadic disaster to ever strike the United States, and I wondered aloud why no one had mentioned that yet. Once again, the Tri-State Tornado was being overlooked.

The panel chastised me, indicating that national radar would without a doubt pick up a storm of the magnitude of the 1925 one, and that it was almost a moot point to have brought it up at all.

I retorted that the Tri-State storm hadn't started at the F5 that it likely eventually became, and had to drop *somewhere*, and wouldn't the folks outside Ellington, Missouri, (where the storm began killing people at an estimated F3 magnitude) have appreciated some kind of warning at all?

Following the hearing, I was descended upon by people who wanted to speak to me about my knowledge of the Tri-State Tornado, many of them meteorologists, some from far-off states (one from Florida) who had come for the Doppler event to witness the arguments and results first-hand, the issue was of that much importance to them. Many of them discussed the outcomes of a possible book project, when I mentioned it; all input was favorable. They were content to settle for the fact that I was still reluctant to write a book, and that I had folders full of information, but all agreed that *something* should be written about the disaster.

At the insistence of those who continued to show interest, not only in the idea of a comprehensive product of what happened that March day 1925, but at the idea that a complete book could be produced and bring attention to the stricken areas even though it was almost 75 years past, I picked up the telephone one night when the calendar changed from June to July 1999 and made a phone call at about 11 p.m. to a lovely motel in Ellington, Missouri.

The result of the ensuing conversation is what you are holding in your hands.

You will be taken on the quest I made, with my children in tow, to track back across the damage path of 219 miles (an astute group of researchers, operating shortly after this book was written [2002], are now saying that number may be as high as 224 miles, as it's possible the storm dropped an additional 4 to 5 miles further to the southwest than previously believed) to chronicle reports from those who lived through the event, and back and forth between 1999-2000 and 1925 in the twinkling of an eye. I have often wondered if how I did what I did would be of interest to those who are purists and devotees of the Tri-State Tornado. I came to the conclusion long ago that perhaps it is….and perhaps there is another disaster out there, one that happened in a young person's backyard, that needs chronicled, and if I can inspire someone else to get out and do what I did, with the obstacles I had, it was worth it to write it in just that way.

This book was written between 1999 and 2001, so the accounts will all begin with roughly the phrases "in (mid-, late, etc.) 1999" and the like. The ages of those interviewed will be given in their account at roughly what they were on the date they were interviewed; but for perspective, I ensured that their ages when the storm struck was also listed within their accounts, and early on within each chapter so the account will show it if the age makes an impact. The average age of survivor I interviewed, when the tornado affected their lives, was eight. Therefore, most of the stories are worded somewhat simply, as they are told from an 8-year-old's point of view. There are photos of these lovely people as they either allowed me to have them, or to take them myself. There is a lack of photos in some locales, where these beautiful elderly people simply didn't want their photos taken. I wish I had been able to capture each and every one of them on film, if for no reason other than my own remembrances of them, for many are now long gone in the time that has passed between the creation of this book, and the actual production, something I feared early on.

The original intent was to publish through a university press. Given the time frame of the finalization of this tome (around 9/11, which should give you an idea of financial opportunities at the university at the time, as those were based upon grants), you might be aware of what pulls the strings of many university presses and what suddenly became 'unavailable' in the weeks and months post-terrorist attack. Self-publishing, once a consideration, became a non-issue with a December 2002 diagnosis of multiple sclerosis and a February 2003 advisement that Systemic Lupus Erythematosus was back after years of remission, along with some more serious and rare conditions with which I suddenly had to contend. I had to put this book on a back burner and concentrate on learning to live with debilitating and disabling conditions, the beginnings of which I had felt the twinges of when the idea of this publication was being formed in early 1999.

Another note: I am not a meteorologist. I have never aspired to be one. So the weather references within this book are merely 'observations': observations made by people mostly who, when their age was in the single digits, experienced something that was a defining moment in their lives and it affected them from thence forward. They aren't meteorologists, either. Between them and I, we're telling you, in layman's terms, what happened and how it looked, sounded, smelled and felt. I wish I could explain better the meteorological and scientific terms the esteemed weathermen at the end of the book use. I am honored and somewhat overwhelmed that they took the time to grant me interviews and bring their expertise to this book. Their work has been hugely important to the overall impact tornadic events have on our lives in the United States.

Never could I have imagined that this project would take on the life of its own that it did. My only regret was that there were so many more survivors out there whom I would have liked to have interviewed, and get their stories, and chronicle for posterity and for generations to come, and due to other responsibilities I was unable to accomplish this. The magnitude of the regret became evident in April of 1999. A survivor, Anna Mary Bilberry, of Crossville, who was twenty at the time of the storm, had been suggested by a friend as one whom I should definitely interview for the story, as she was still quite cognizant and could speak coherently about the event and the impact it had upon her family (she had lost a younger brother in the storm).

Miss Bilberry died in a nursing home in April 1999 before I could get around to contacting her.

My thanks are to the 54 men and women interviewed from the original inception of my informal research, (February 1999) to the later, more difficult-to-find interviewees (the McClurkin sisters in Princeton in December 1999, and Lee Rauscher in May 2000). Their information and some never-before-published photos have been invaluable. Thanks to Rosetta Adams, Ralph Gates, the West Frankfort Genealogical Society, Old National Bank in DeSoto, Illinois, the incomparable and beautiful Lola Jones and to the many others who donated photos for reproduction, such as the McClurkin sisters in Princeton, who told me that the photo of their relatives and neighbor standing on their ruined property had been stowed away for years and hadn't seen the light of day until I came to interview them; they trusted me with the photo, and I appreciate that beyond words. Thanks to Margaret Russell and Mary Belle Melvin in Murphysboro, who, with their own hands several years ago, compiled accounts from that town. Thanks also to the National Funeral Directors' Association in Brookfield, Wisconsin, who went the extra mile to ensure that I had the somewhat obscure accounts of their entities' efforts, post-storm. Extra super thanks to Stan Changnon, Ernie Kern, and the most recently-contacted weatherman, Charles Doswell, for the input about the technical aspects of storms and meteorology. Thanks to Tara Eveland Photography for use of the picture of me and one of my granddaughters for the back cover.

Above all thanks to my darling family, which has expanded considerably since the three hyper little kids everyone thought were so adorable (because, face it, they *were*) were dragged with me all over three states in the summer heat and winter rain to gather the information so important to this book.

And to my husband, who has put up with so much…helped me learn to walk when MS wanted to keep me from it…helped me fight the bad times and learn to enjoy the good times…has proven to be that rocky crag that a person can run to when the storms of life strike: I love you. Mere words cannot say it. Only a lifetime can show it. I give you mine.

Angela Mason

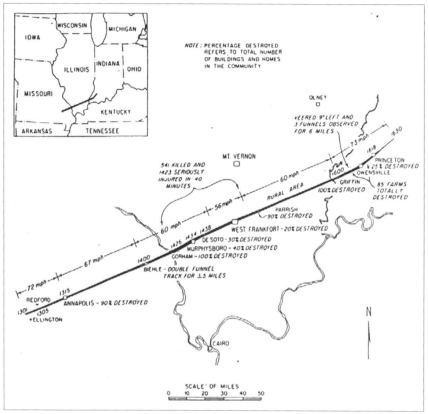

This is the long-accepted diagram, created by Stanley Changnon and John W. Wilson, showing the track of the Tri-State Tornado. Recent evidence uncovered by Dr. Charles A. Doswell III, however, may mean an additional four to five miles, and perhaps more, could become part of the tornado's tract at the outset, prior to Ellington, Missouri. This theory is explored in Chapter 18. The Tri-State Tornado remains the single most devastating tornadic storm in Americn history, and hholds the record for number of deaths (695), number of injuries (2,027), highest estimated wind speeds (over 318 miles per hour), damage path width (estimated 1.5 miles in Hamilton County, Illinois), path of destruction (219 miles at the present) and time on the ground (almost three hours 30 minutes) for a single tornadic storm. It is hoped that no tornado will ever again wreak the kind of devastation the Tri-State Tornado did, but if ones does, this book may be a valuable guide if for no other reason than for the examples of the kinds of resiliencey the human spirit can display, and ultimately prevail over any adversity. The maps are from the circular "Illinois Tornadoes" by Changnon and Wilson published by the Illinois State Water Survey in Urbana, Illinois.

Part One: The Sound and the Fury

Missouri

Chapter One

Ellington/Redford, Missouri

March 18, 1925
between Redford and Ellington, Missouri
11:50 a.m.

The horse Samuel Flowers regularly used to ride the rough terrain every couple of days or so to his brother's house was all geared up, but Sam was worried about the saddle.

The girth strap was rather worn; its cinch was frayed in the way only leather can be when being sweat upon and constantly rubbed to and fro by the motion of horseback riding. It had made a whispery sound when Sam had pulled the band through, not the usual healthy squeak of a leather strap being locked in place by the cinch rings. Truth be told, the entire saddle was going to hell; saddles were generally made to last the lifetime of a horse and then some, but this one had seen several lifetimes, including not only Sam's horse, but also an old mare that belonged to John Flowers, whom Sam was going to visit about seven miles away down in Dry Spring.

John had a young'un down with a sickness that had puzzled the normally ailment-wise old woman midwife who lived in the local holler. She had been called in two days ago. Now the family was at its wit's end over the child. Sam felt pressed upon to go and receive a status report on the little boy, and bring back such news to his own family so that if there was anything more to be done, perhaps it could be discussed and arranged. Options were to take the child on over to Perryville to a very good doctor who had recently set up shop there, a long and arduous trip to the west, but one that might be worth the effort; or maybe hold a prayer vigil to ensure a cure beyond what was normal for an illness that itself appeared to be somewhat beyond what was normal. Maybe the family would seek a little bit of both.

Whatever the choice, Sam would be involved in it, for he had the sturdiest mare and the friends down in Ellington to whom his trusty mare would take him. And those friends had a good automobile that could make the trip on up to Perryville if necessary.

The morning itself had been misleading for mid-March. There wasn't the slightest hint of a breeze down in the holler near Logan's Creek, where the Flowers had a homestead. There was a bit too much warmth in the air for March 18, and this always rattled Sam, who was an early gardener and knew better than to trust Missouri's spring weather. But there had been years where there was a lot of odd weather to be seen when the winds of March were this warm—rain being the greatest hazard. The runoff in the hills and hollers of the Missouri Ozarks was notorious for being unpredictable. Some rains seemed to soak the ground and nurture it, while other rains were rejected by the limestone and shale formations that were the underlying strata of the fine loam soil. Rejected rains wreaked havoc on the sparse patches of farmland and sustenance locals cultivated. A heavy rain this early could mean replanting and a diminished harvest of crops many families and livestock needed for their very existence.

Worn saddle aside, Sam Flowers climbed up and took the reins, and his fine mare followed its lead, trodding down the rutted track that served as a road for those venturing into the hollers through which Blair's Creek and Logan's Creek snaked their way. A little dog, a terrier mix with black-and-white colorings on its short hair, ran yipping after the horse.

"C'mon, then," Sam yelped. The little dog jumped and barked happily at its master's voice coming to him from high above; "let's go if'n ya have to."

The dog fell in behind the mare's stride and let go a series of sneezes. Man's best friend shook his head. His ears flopped in a hapless manner. He didn't like the feel in the air; this was, in fact, the primary reason why the dog opted to follow his master as opposed to staying behind at the clapboard house with the many other members of the family. Though the dog couldn't voice it in any manner other than to rattle its ears until they popped along the side of its head, the canine felt it and put forth a mighty effort to stay within barking distance of the horse. Something told him he didn't want to be far from his master.

July 2, 1999
Fairfield, Illinois
5:45 p.m.

It was the beginning of July, 1999. It was hot, the hottest weekend of the summer so far, and this was the one we chose to take off and take a road trip back through time.

The motivation of myself and my three children, ages 11, 8 and 3, was simple: It was a holiday weekend, the Fourth of July to be exact, and everybody's going to be around the old home place because that's the kind of holiday it was—stay at home, barbecue, have little family reunions or block parties or church get-togethers. Finding people to talk to under these circumstances, I figured, would be easy. Therefore, even though it was hot to the sulfuric degree, we were going on this trek in an air-conditioned vehicle to see if we could catch some of these folks at home. I was seeking survivors or relatives of survivors or those who knew survivors of the worst weather disaster ever to hit the central United States.

Nobody really thinks of spring storms when the temperature hovers in the upper 90s. In fact, a thunder-boomer would be welcome relief in this insufferable heat. But there was not a cloud in the sky on the evening we left our comfortable, air-conditioned abode on the eastern side of Illinois heading west. The weather reports had been warning all week that we were in for some serious sun and humidity.

Spring storms? That was a few months ago, an eternity from when the dog days drag on endlessly in the summer. Yes, a big blow came up in early June and knocked over a few trees in the backyard. Sure, my neighbors down the way had a sidewalk uprooted and lost the underpinning on their mobile home. "Don't like those spring storms, but, ya gotta live with 'em," they'd say.

These same folks had only a vague idea of the type of storm I was chasing almost seventy-five years after the fact. These folks who bemoaned the blow that came through in June lived north of the path of destruction left by that earlier killing force, a force I was now seeking tales about. They knew a few of the relatives of folks down south affected by the tragedy; they heard from some hearty souls who went to view or to help about just what a "catastrophe" and a shame it really was, and what the size of the twister or twisters had to be in order to create such devastation on that level.

Their elder relatives knew devastation, many decades ago, and they died after passing the information to their children and grandchildren. Those, in turn, grew and briefly mentioned the same story to their children but left the grandchildren out of the loop because it was "a little too much for them to handle at such a young age."

So those folks, one of whom was I, didn't find out about such a monumental event until they were a little older. I had heard stories and had asked around to see just what the reality of it all was.

Others simply didn't care, as they grew up, because they didn't want to consider something as hideous as a bad, a very bad, storm, something that, in the day and age of control freaks and of making your own destiny, was a very undesirable thought…because one couldn't control the weather.

Our destination this evening, the 2nd day of July 1999, was Ellington, Missouri. We didn't know how long it was going to take to get there; all we knew was that it took about two and a half hours to reach that western edge of Illinois.

Chester, Illinois, was where we crossed the mighty Mississippi and entered the eastern part of the next state, Missouri. As we drove in the fading sunset, the clear, gorgeous sky made it difficult to imagine, as we passed that point on the Illinois part of Interstate 57 where the tornado of 1925 crossed over the flat farmland, just what the sky looked like on that day. I wondered what the surroundings had been and where the people had stood wondering just what the hell was going on as this murderous beast of a storm barreled out of the southwest. Oddly enough, our route, in order to reach a crossing point into Missouri without going too far out of the way north (St. Louis) or south (Cape Girardeau) took us along the very path the twister took. It was not a feeling of terror or morbidity that overtook me as I drove along the tornado's path (backtracking the direction the twister took), rather, it was sheer awe, wonder, and the utmost respect for one of the most unpredictable and unbelievable forces on the planet.

*   *   *

We arrived in Ellington, after a harrowing and accidental tour of the foothills of the Ozarks (and just south of the highest point in Missouri), a little after one a.m. It took us five hours of drive time—two were spent backtracking our backtracking and getting lost, until we stumbled upon Ellington.

Ellington, Missouri was a very sleepy backwoods town nestled like a jewel in a valley of the Ozark foothills. Our motel room awaited us. So did a noisy gang of Pre-Fourth of July revelers, with a cooler full of beer and a truckbed full of those partaking in spirits.

"For me?" I shouted, putting my hands to my face in mock surprise as I emerged from the SUV, rattled from the road of ups and downs and all-arounds in the mountains. "You guys must have known we were coming!!! Really, you shouldn't have!!!"

The revelers greeted us with big grins and open arms, noting how funny it was for a harrowed-looking mom and her three sleepy kids to pull into the motel parking lot at one a.m. In my typical way, however, I ignored them and instead of a beer, I opted for the room and the chance for the kids to sleep before the events of the next three days, and consequently, the next nearly two years, unfolded.

We were going to need it.

Just north and east of this little burg, with its beautiful view of the hills and rock crags all around it, sat the location where the first of 695 was killed when the tornado began its high-stakes roll of the dice across the Ozark Mountains of Reynolds County.

My intent with the rising of the sun was to discover who this person, from reports a farmer, was, and hopefully, where exactly he was when the tornado reached out of the heavens and claimed him for its first victim. With success, I hoped I might even find someone who knew him…what he was like, how old he was, was he married with children and did he always take his Saturday night bath or did he blow it off once in awhile. It was an ambition of epic proportions, like finding a needle in a haystack. So I thought, as I drifted off to sleep in the comfort of the fine motel room I'd secured just over 24 hours before.

Little did I know what the next 24 would bring.

March 18, 1925
Dry Spring, Missouri
12:30 p.m.

As it had turned out, John's baby boy was going to be just fine. The fever that had held the child in its grip had broken, leaving relieved parents, siblings and others to feel gratitude to the Lord for watching over them.

"He works in mysterious ways," John was heard to comment on his God as Sam mounted the mare for the ride home.

A low rumble of thunder, more felt than heard, caught the attention of both men.

"Maybe He's speaking to us," Sam muttered, not convinced. "Telling us this might just have been a sick boy whose time it was to get better."

John laughed a derisive snort toward his younger brother. He knew Sam was not the most religious of men; likely, the most religious his brother had ever been was during a stint in the Jefferson City Prison. Sam's eyes had been opened over those few years; the crime he had committed had been attempted murder, the conviction decisive, punishment quick, the jail time more of a following of "old rules" than of punishment. In fact, the crime had been committed more by a following of those "old rules" than out of malice, in the ways of the hills people. When a man had a grudge against somebody, he didn't walk around it. He met it head on. And Sam had met with that somebody head on, and had tried to put a bullet to the head of that man. Dealing with that fact generally made men hard or softened them to a higher power that they allow to take hold of their lives a little more once jail time had been served.

Sam hadn't been out of jail too long, a little less than two years, in fact. He'd been more of the latter; his faith hadn't been restored fully, but what was left had been refined into a healthy respect for the Creator, and not much more. John saw his brother glance over his shoulder with hesitation as a second low rumble issued a report across the hills that ringed the Ellington area outside Dry Spring.

"Blue skies," Sam muttered wonderingly as his eyes searched the azure above him. "Must be quite a ways off."

John followed Sam's gaze. His eyes narrowed, and he felt something kick over inside, something that his old momma would've called intuitin', and he placed one gentle hand upon the flank of Sam's mare.

"Why don't ya stay here," John offered. The tone in his voice was more of insistence than of invitation. "It's still such a piece down there to the holler. It might come up a storm afore ya make it back."

John Flowers moved around beside the mare's head and took hold of the bridle. From here he looked up to his brother and once again voiced his concern.

"Stay here, Sam," he repeated.

Sam Flowers glanced back up at the stunning March sky and shook his head once, briefly.

"Gotta get back home," Sam said as he lifted the reins.

The mare tossed her mane, her forelock flipped and she rolled her eyes at the little dog that had appeared from an outbuilding to join its master.

"Got baby Sam to look after," Sam spoke of his year-old son by way of excuse.

And without further conversation, Sam Flowers took out down the dirt road away from his brother John's place in Dry Spring, and headed for the hard road that lead over toward the holler where his family awaited him.

The little terrier mix threw one glance over its shoulder at the brother Sam was leaving behind, as if to invite him once more to insist that Sam stay. But John merely frowned and bit the inside of his lip, a habit he'd never been able to break. He walked back the short distance to his own home and to his family waiting inside.

To accompany the men as they went their separate ways, thunder grumbled again from an unspecified direction.

July 2, 1999
Ellington, Missouri
12:50 p.m.

At the Queensway Restaurant in Ellington, I discovered that everyone in town knew the story of the tornado.

Not one of them, however, confirmed that it struck Ellington but rather an area outside of it, to the north and east, near a little burg called Redford. In order to follow up on the lead, I took the advice of some of the locals and went to talk to Marvin and Lucille Fox, who lived up in the hills outside Ellington and in the Blair's Creek area, not too far from where Sam Flowers had once had his homestead.

Their homestead wasn't much; that's putting it simply and mildly. The house had obviously stood there for many a decade, balanced, it seemed, on top of a little rise known as a 'hill' as opposed to a 'holler.' To get to it, one was best outfitted with a compass, or at least a keen sense of direction once directions were given. The locals were able to deftly give highway instructions as in 'Take K road to whur it runs inta CC, an' cross 21 afore ya get to the iron bridge; if ya cross the bridge, you've gone too fur.'

That was a bit too complex for mountain terrain; in order to find the Fox residence, I was best advised to seek out and find the first 'hard road' that turned off State Highway 21 as it snaked its way north into the mountains.

The 'hard road' gave way to gravel shortly after the turn into the Fox's drive. Winding around to the back of the house, we could see it was lined with pecan, apple and peach trees, all of it bearing fruit, to the delight of my children. Up ahead, and forming the periphery of the property before it dipped into a holler, was a row of thornless blackberries, of which Lucille Fox was understandably proud, as they were beautifully maintained.

Fuzzy, playful kittens of all colors and designs prowled and leapt about the underpinnings of the clapboard house in which the Foxes had lived all their married lives. Marvin, 83, and Lucille 71, were sitting outside on the back porch (indeed, the only porch on the house). There was no air conditioning.

"We've never needed it," Marvin insisted when I asked him about it.

I sat down next to them in the shade of giant maple trees. He was probably right. The breeze that drifted steadily through the woods and across the hilltop was adequate, even though the mercury was reading 94 an hour after noon. In a house as well constructed (though aged) as theirs, I could understand how air conditioning could still be considered a luxury by these fine hills people and not a necessity. Except for a few gnats, the side yard was quite comfortable.

Marvin knew the Flowers; seemed he was related to them by proxy, as his sister, Edith Fox, had married a Flowers, one of Sam's nine children. But Marvin, who was hard of hearing and not in the best of health, didn't say much about the tornado nor the Flowers. He left that up to his wife, Lucille, who knew the story well and proceeded with what, to their family, was a tale that was known and had been told and retold for the past seven and a half decades.

March 18, 1925
between Ellington and Redford, Missouri
12:45 p.m.

The mare was growing skittish.

"Hesh, Babe," Sam Flowers told her, calling her by the only name he or anyone in his family had ever given her.

Babe gingerly stepped over the hard-packed dirt that grew rutted from rain

Lucille (Gore) and Marvin Fox at their home outside Ellington, Mo., July 1999.

and wagon wheels as it approached Blair's Creek. Oddly enough, for every few steps she'd take forward, she'd do a little jog to either the left or right, as if to indicate that she wanted to backtrack—Sam wouldn't let her.

"C'mon, now!" he said, and noted with a lurch in his stomach that his words had come out in a growl.

He knew a horse could detect how its rider was reacting to circumstances affecting him; and right now Sam was growing concerned. The sky had taken on a steel-blue cast to the southwest, which was directly over Sam's left shoulder as he headed north and east on the wagon road. Not only did it not look right, but it didn't feel right either. Sam found himself gulping or gasping more than one breath in a row. He thought fleetingly that maybe it wasn't the air; maybe it was just him, his 49-year-old lungs perplexed by his reaction to the odd tint of the sky behind him.

But the mare knew it too. She began adding an erratic pattern of snorting and blowing to her sidestepping, an accompaniment that gave Sam the jitters. She emphasized her discomfort with several tosses of the head, her mane and forelock flying as a result. There was no denying it. The horse was getting spooked.

Sam reached up and patted her neck firmly. As he did he noticed a thick trickle of sweat crawl from his armpit, running down his ribcage before meeting up with and soaking into his broadcloth shirt. Sam jerked his arm back, the sweat bringing to mind a fat spider that crawls lazily across the landscape of skin before biting once, twice, three times. But there was no spider. There was only the trepidation of a force that no human could control and further, could not endure without some innate response.

The mare didn't like Sam's reaction one bit, and literally jumped along with him.

"Easy, Babe," Sam tried once again, attempting soothing but conveying more apprehension than anything else.

The terrier set up a yapping that rattled Sam's nerves instantly.

"Stop it!!" he hollered more in desperation than insistence.

Babe could take no more.

At the sound of Sam's voice, she panicked and reacted as if she were being pursued by the devil himself. Her head suddenly stretched forward, her neck craned against the reins, and she broke into a dead run, a difficult and somewhat unwise move on the rutted wagon road.

He let her run. Sam knew Babe knew the way home, so he held the reins loosely and afforded himself a backward glance over his left shoulder.

The sky stared back at the man with its icy steel cast, as the heavens directly above the hills seemed to become impenetrable. It was as if there were a blanket about to descend and conceal the just-budding treetops, perhaps even the trees themselves. Sam couldn't see whether or not they had begun to bend, a telltale sign that a storm was swooping down into the holler. All he could truly ascertain was that a tremendous rumbling, almost a growl, was building just beyond the hills, perhaps down in Ellington. He hoped his brother had gotten himself and his family to safety.

Sam hoped he could do the same for himself.

Despite fighting valiantly to stay up with the horse, the little dog was being left in the dust of the wagon road. For the canine, keeping up was an impossibility. Babe had urged herself from a run into a full gallop. Sam did not look down at the wheel ruts beneath his mare's somewhat haphazardly shod hooves. He hoped only that she had enough confidence in herself to obtain the footing it was going to require to maneuver across the hard-pack until it gave way to loose rock just before his homestead.

He did, however, chance one last look back over his shoulder, to determine whether the rumble, something more akin to a locomotive passing by in close proximity, such as he'd heard frequently down in Ellington, was descending upon the tree line at the head of Blair's Creek. It was as if it cheated in an effort to ensure there was no warning him of its presence.

Sam Flowers gripped the saddle horn in awe and terror, losing the reins entirely. As they flapped haplessly to the sides of the mare, and dangerously near her feet, Sam watched, horrified, unable to do anything but moan softly as he began to tremble. A shudder welled from deep within and grabbed hold of Sam's heart as the trees behind him were uprooted and tossed about, possibly hundreds of feet, into the air.

It was 1:01 p.m., March 18, 1925. The shadow of night had descended upon the sleepy mountains of southeastern Missouri from a fabled blue afternoon sky. A killer screamed toward Sam Flowers, and there was nothing in heaven, hell or on earth that could stop it now that it had begun.

July 2, 1999
Residence of Marvin and Lucille Fox
outside Ellington, Missouri
1:15 p.m.

Walter Fox felt compelled to speak.

"My parents stood out in the yard and watched it go over," he proclaimed, with not a note in his voice to indicate that his parents recognized the foolishness of that act. Yet, that was just what Roy and Mary Fox did, and from their vantage point at the head of Blair's Creek they could see the apocalyptic storm sweep down into the hollow and begin its devastation as it moved away from them, to the north and west.

The wind was so powerful, however, that it did not leave the Fox property unscathed.

"It moved the house and barn around on their foundations," Walter said. His parents, he remarked, were likely relieved to be outside of the house at that point, as, from inside, there would be no telling what was actually happening.

However, the buildings were shifted as opposed to demolished, due, no doubt, to being closer to the powerful vortex of the storm as it began building in the hollow, instead of falling victim to straight-line winds.

"Storms can get bad down here," Walter observed calmly. "There's hills all around. Storms are hard to see; they boil up. They can be on ya all at once."

According to Walter, his parents noted the storm started at the head of Blair's Creek, in a valley, and then moved down into the creek bed itself, following it closely. The hills and hollows didn't deter the brutal weather in the slightest. In fact, commented Walter, "Storms raise and lower here in the hills. It helps them build."

The fact was the storm took on an initial ferocity in southeastern Missouri that was not rivaled until it met its end in Indiana several hours later.

On an eerily straight heading of N 69° E degrees, the killer storm began with a forward moving speed of 72 miles per hour, and winds estimated to be in excess of 250 miles per hour, according to data obtained in the days and weeks following analysis by engineers and meteorologists alike. Emerging and continuing research beyond the year 2000 has indicated that the touchdown in the area between Ellington and Redford may actually have been 15 additional miles *before* the originally-pinpointed spot (south and west of Ellington), causing the length of the continuous track, previously believed to be 219 uninterrupted miles, to now possibly be 234, the longest tornadic track on record.

At its terrific rate of speed, the Missouri tornado blasted through the foothills of the Ozarks plowing a path between Ellington, just to the south, and Redford, to the north and east. Those in the path of this storm who lived to tell of its fury had described their accounts to Lucille Fox:

"My aunt and uncle, Carrie and Johnny Chitwood, lived a little further up Blair's Creek. The storm blew away half their house and left the other half standing," Lucille began.

"My cousin, Rosie Chitwood Dillard, had her two little kids with her that afternoon, and she was goin' across a field when she heard it and saw it comin'. So she ran to the edge of the field, in a tree line, and lay down with those two young'uns underneath her! And she was a big woman. But she grabbed hold of a hickory sapling and she held on while that storm went over. She was able to stay down and those little kids were able to stay under her. But she said that hickory sapling 'bout beat her to death whipping back and forth in the wind."

The memories didn't stop there for Lucille.

"My mom was really affected by that storm," Lucille admitted, a note of sadness touching her voice at the memory. "It scared her so that every time it'd come up a bad storm after that, if it was the middle of the night, Mom would get up and get dressed. She just wanted to be ready."

Lucille had a considerable amount of family in the hills, and many of these kinfolk had plenty to tell about the storm that happened years before Lucille was even born. One storyteller was her father, Henry Gore.

"Dad always told about how he'd been riding a bay mare that he really trusted. He'd had her for years. And he was riding through the holler that day and the horse was right with him on trying to get out of the way of that storm. Dad said he just leaned over her and let her have the reins. She went really good. She leaped over fallen trees, and dodged trees lapping over in the path, or through treetops if they were flat laying down in the wind. I don't see how he did it. But Dad always said that if it hadn't been for her, he would never have made it home."

And what of the other mare rider, Sam Flowers?

Walter Fox explained.

"His horse came in without him, back to his house. And a bunch of folks went out there looking for him."

Jim Flowers, whose father, Clifford, was one of Sam Flowers' nine children, was a kind gentleman and an avid rock collector, who lived in the hills on the north edge of the Ellington city limits. Jim added to the account of what happened to his grandfather, which was vivid enough to Jim's dad Clifford at the time, as he was 21 when Sam Flowers became the first victim of the tornado.

"A man named Irvin Barnes found him," Jim said. "The horse had gone on home but that little dog stayed with Sam and set up a barking and that's how they found him. My dad said the tornado either blew him off the horse or he got off and hung on to a tree and it fell on him."

Walter Fox elaborated a bit further, as he was actually present at the time of the disaster.

"Sam's head was either between two trees or a fork in a tree," Walter added categorically. "Either way, his head was mashed flat.

"They had to have a closed casket at the wake."

*   *   *

Samuel Flowers had just observed his 49th birthday, March 16, two days before. He left nine children, including a year-old infant son, Sam, Jr., and a wife, Mary Adeline Flowers, "a little bitty ol' girl, not over 100 pounds, a beautiful lady," said Jim Flowers.

Sam Flowers was a farmer, as Jim Flowers put it, "like everyone was at the time. Everybody was either a farmer or a timber worker."

Sam and Mary Adaline Flowers gravestone located in Redford Cemetery.

Although Sam had spent time in prison for attempted murder, he was not spoken ill of in the communities between Ellington and Redford. Rather, he was remembered by the old-timers, those eighty or older who might still be able to recall him, as a decent feller who thought a lot of his family and who worked hard to eke out a living in the rocky soil of southeastern Missouri. Sam was buried in the Redford Cemetery, his wife Mary Adeline later interred next to him. Theirs was the first grave to the right past the entrance gate. Mary Adeline Flowers died in 1963, having lived out her years in St. Louis following the 1925 tornado.

She never remarried after Sam's death.

And on March 18, 1925, the tornado, blasting through one of America's most beautiful and scenic landscapes, left the Redford/Ellington area and barreled on toward Annapolis, Missouri. The toll thus far was one dead; two injured (this was three percent of the total population of Redford dead and injured in the storm); property losses: $5,000 U.S. dollars circa 1925. These were the initial losses and there would be more—much more.

Chapter Two

Annapolis, Missouri

The view from the winding roads as we made our exit from Ellington was phenomenal, even in the oppressive heat of summer, or perhaps because of the oppressive heat. The trees seemed so green, the sky so strikingly blue, that there appeared to be no way a killer like the tornado we were tracking could have wound its hideous path of mass destruction through here.

The rivers and creekbeds of southeastern Missouri were fascinating to me as a child of Southern Illinois. In Illinois, riverbeds were mud. It was fine to catch fish in, to boat upon, but to swim in was atrocious because the mud consisted of fine silt and good Illinois topsoil that washed off in heavy rains and floods and created a silky muck that bordered on slime. This was generally true of the lining of every creek and river in at least the southern third of Illinois.

But in southeastern Missouri, where the underlying foundation of the soil everywhere was composed of different kinds of rock, quite the opposite was true. Every tree and blade of grass and tuft of moss that clung to the soil did so tenaciously, because that soil clung with equal tenacity to shale, sandstone, and limestone. In the river and creek beds, the soil was consistently washed away and what was left was a stunning layer of stones and pebbles. Most were sand colored, some white, some with streaks of gray that indicated a high iron content.

Iron and other ores were the reason why Annapolis came into being. From Redford due northeast on K Highway, any map would indicate that it was almost a straight shot to Annapolis. Under no circumstances be fooled. There was nothing straight about the mountain highways (the locals refused to call these ridges in the midst of the United States 'mountains,' they preferred the term 'hills' instead, although the Redford/Annapolis area was just south of Taum Sauk Mountain, elevation 1,772 feet, the highest point in Missouri.) The road, a well-maintained, high-grade blacktop, wound perilously in and out, up and down and around, from the little four-way stop 'town' known as Redford. The road crossed deep gorges and high cliff-like mountainsides where the hills had to be blasted away in order to create a path. On occasion, the cliffsides occurred naturally, and the view then, as throughout the region, was truly breathtaking.

The roads continued on past forests and pastures where cattle, horses, mules and the occasional goat grazed, past farms and isolated houses, (both new and stunning, and old and decrepit). Cemeteries and gardens lay peacefully, until quite suddenly, a rise in the road appeared ahead and as it curved to the right ever so slightly, there sat Annapolis.

Arriving in Annapolis from the south, it was noticeable that almost the entire right-hand (east) side of the town was a mine. There were mine tipples, loaders, augers, and a host of other equipment that only those who were involved in the mining business would know the nature of. The mines were obvious in the immediate area of the hills surrounding the town; they were also within the county of Iron, as it was called due to the enormous deposits of the ore found in this part of Missouri. Directly north of Annapolis were villages and towns with names like Chloride, Ironton, and Graniteville. There was also at one time a tiny mining community by the name of Leadana, derived from the nearby Annapolis and from the ore that the mines produced.

Leadana was developed, in the early part of the twentieth century, by miners for miners. There had been a grocery and other modes of trade for those miners residing there with their families. As well, there'd been a small schoolhouse set up by the mine company for the children of the miners. There was even at one time a doctor's office to treat the resident mine employees and dependents. The workers and their families had resided there in somewhat substandard housing, as the information available these days related. The miners used to work long, hard hours and got to see their families in the light of day for only a few hours at a time. It was a hell of a way to make a living.

However, it was an incredible way for the miners to have lived, as they did, through the worst weather disaster ever to strike the area, as many of the miners might not have survived the day had they not been underground, hard at work in their chosen vocation. The death toll, being at a minimum in the Annapolis area, would no doubt have been higher had the storm struck in the hours when the miners weren't working. There was a good possibility that it would have permanently affected the town's population.

The Annapolis at the turn of the 21st century had a population of 363. One of those 363 was Mrs. Clara Brown, whose family in 1925 lived on a farm four miles west of Annapolis.

"It was like a curtain came down over the sun"—Clara Brown, age 16 on March 18, 1925

Clara Brown was a mere child of sixteen at the time, but in 1925, sixteen was an age when young girls finished their education and set out for a life of their own with husbands if they so chose.

This was Clara's intent, to do as her two sisters had already done and start families with their husbands, building homes in the city limits of Annapolis. Both brothers-in-law and her fiancé, Carl Brown, were miners who worked in the Leadana area.

From her cozy brick home in Annapolis, the town where she had spent all her life, Clara, at the age of 90 in the summer of 1999, recalled with devastating clarity the events of March 18, 1925.

Clara remembered it was a beautiful day, warmer than most mid-March days in southeastern Missouri.

But for her family, it was a day just like any other.

"My dad and mom and I were out at the farm back then, and that day we were all sitting inside, just after lunch. Dad was reading the newspaper and I remember he looked up and said something about seeing a 'very black cloud.'"

It was one-fifteen in the afternoon.

It had taken the southeastern Missouri mountain storm just fourteen minutes from its rapid development more than twenty miles away, traveling over incredibly hilly terrain, to reach the burg of Annapolis—with a forward moving speed since calculated at 72 miles per hour.

"Dad said, 'Oh, it's storming,'" Clara recalled, not at all understating the gravity of the statement in its simplicity. Instead, the words her father spoke, issuing from Clara, carried with them all the dreadful seriousness of a storm the magnitude of the one that was approaching the town, outside of which their farmhouse sat. Clara provided a fine replication of her father's tone. That unpretentious sentence was turned deadly by that tone.

Their particular farmhouse escaped serious damage, as it was to the south and west of the heading of the storm. But Clara and her family watched in frozen amazement as the black, monstrous clouds skipped over the hills and swept down into Annapolis with an undeniable, evil fury.

"We'd never seen anything like it," Clara remembered softly. "The one thing I can say about it is that it went through really quick."

Clara nodded at this point, obviously in agreement with herself as she made a rather astute observation about the storm.

"It sounded simply like a train coming through," she mused.

"It was like a curtain came down over the sun."

> *"It was like this…It was dark and it was light and it was over"—Alice 'Peachy' Jones, age 10 on March 18, 1925*

Alice Berniece Smith Jones carried a lot of names for such a little lady, but there was one more, a nickname, added to that list. Born Oct. 15, 1914, she was the apple of her father's eye—or, as the case may be, the 'peach.'

"My dad owned a sawmill," she said, "and for years he would come back to the house and feed the men who worked for him. When I was a baby and Dad was home for dinner with the men, I might be a little fussy so he would pick me up to keep me quiet while everyone else had their dinner. And he sang a song to me that went 'You be the peaches, I'll be the cream.'"

Alice Berniece Smith Jones smiled.

"I was the peaches," she said, hence the nickname she acquired from Dad that stuck with her to this day, almost 85 years later in the hot summer of 1999—Peachy.

She, like Clara, was a mere child, not quite eleven on March 18, 1925, and was in a school that was called, simply, Annapolis School.

"It was a four-room brick school," she said, "two downstairs and two upstairs. I was in the downstairs room on the left, the one for the second grade.

"We were getting called in from noon recess, so I know that this tornado had to have hit after one o'clock," Peachy noted. "The kids were screaming and hollering when they saw that black cloud and heard that awful noise. So the teacher got us and gathered us up around the desk, near the inside wall, like a bunch of chickens."

There were quite a bunch of chickens in that schoolroom, 25 to be exact, and their cries and wails could barely be heard above the racket as the storm roared overhead. Traveling at the end of its 72 mile per hour forward moving

Alice Berneice "Peachy" Smith Jones at her home in July 1999.

speed, it showed no signs of slowing down, although that's just what it was about to do. It was as if the killer storm made a conscious decision to take more time with its next victim. But moving at 72 miles per hour made this simple impression on young Peachy's mind.

"It was like this," she indicated and ticked off one finger per conjunctive phrase as she said, "It was dark and it was light and it was over."

Peachy's measurements were accurate, for in about the time it took for her to speak those words, that would have been the amount of time of the duration of the storm as it passed over Annapolis School at 72 miles per hour forward moving speed.

What did a storm moving that rapidly and ferociously leave behind?

"Everything was just flat," Peachy remarked. "There were a couple of big boys upstairs, and their mom was sickly and they got scared for her."

What weren't flat in the schoolhouse were the doorframes, which still stood with the doors practically intact. Amusingly, these 'big boys' pushed open the doors, as if that were the only escape from the rubble that the brick schoolhouse had become in the aftermath of the powerful storm.

"And when they got out, all the kids then poured out" of the rubble, according to Peachy, and all began running into town.

*   *   *

Clara Brown's two sisters were living in town in 1925, and Clara's parents loaded up with her in the family's vehicle and attempted to get into Annapolis, but were stunned—and their progress halted—by the amount of debris blocking their passage. Not to be deterred, they got out of their vehicle and walked into town, like so many others were going to be doing.

Amazingly, both sisters, one of whom had small children, were found to be okay.

"My sister, Shirley Johnson, had heard the roar before she saw what was coming, and she knew, so she had an iron bedstead and she went and put those kids behind the iron bars, and that, with them being up against the wall and those bars in front of them, was what saved them," Clara said. "My other sister, Macey Slusher, had had a fine house in town, and it was blown down, with all the furniture damaged."

Clara and her family continued on into town, more in a dazed manner, she noted, than with any specific purpose. The wreckage was almost beyond comprehension.

"Outside of town, we could see where the storm, if it was tornadoes—'cause we didn't know for sure at the time—had cut a swath of my dad's good timber, across Bear Branch, as wide as a big house.

"And when we came into town...oh, we found such a terrible, terrible thing...children crying, they couldn't find their mommas..."

Clara took a moment and shifted thoughts toward her own family instead of the memory of injured children who thought, and fairly so, that the world had come to an end.

"We helped get our folks home, helped take care of our family's kids. Because, it was such chaos. The people who had people went to their folks—when they finally found them."

Those in the mines in the Leadana area, while spared the sight and sound of the storm, were in no way spared its eventual wrath, however.

"The storm tore the mines and machinery right up. There weren't very many men above ground, but my cousin's husband was killed there, when the tornado went through Mine's Hollow," Clara said. Mine's Hollow was what the Annapolis folk called the area of Leadana.

"There were up to 450 underground in the lead mine," she pointed out.

Those men climbed to the surface only to find their mine operation, grocer's, doctor's, and clapboard homes destroyed.

Women had escaped serious injury by taking cover; children had been in school. The latter structures were somewhat sturdier than the clapboard houses.

Seventy-five years later, "there's nothing there now," Miss Clara Brown related of the 'town' of Leadana. "They went ahead and rebuilt the mine offices and equipment and it ran for a little while after the storm, but after a time, it shut down."

*   *   *

Peachy Smith ran into town as fast as her ten-year-old little legs could carry her.

"I saw folks I knew," she said, "and they said I ought to wait right there (in town), but I knew I had my mom and brothers in the house, and I wanted to get to them."

When she reached her house, which was on the north part of town and up a little rise, she was stricken by what she beheld. Basically, she saw a house torn inside out.

"That tornado had blew the roof off and took it away, and it lifted the walls and dumped everything outside," she remembered.

Peachy had a fifteen-month-old baby brother, Archie Ray Smith, and an older brother whom Mom had ordered to stay put in the house after he had come home from school for lunch.

"He'd got home just after noon," Peachy said, "and about the time he was supposed to go back, Mom had seen the clouds coming and told him to stay. He was blown out into a pile of wood. He wasn't really hurt, just banged up."

Peachy's mother did not fare as well.

"Mom had a big timber fall across her back," Peachy said. "Now, I didn't see it at the time after it happened, but they told me that it tore a big hole in her back."

In an attempt to shield their daughter from the horrors of the type of wounds the storm dealt, her mother's injury was never revealed in full to her by her parents. Peachy did not have accurate knowledge of it even 75 years later.

"I didn't see anyone else's injuries, and I didn't know of anyone injured badly except for mother. I didn't see the place on her back. It took months for it to clear up."

Baby brother Archie was saved by mother's instincts.

"He was okay in Mother's arms," Peachy noted. "Mom said when the wind threw her through the house, she hung on. She always said, 'It's a wonder I didn't squeeze him to death.'"

Peachy's father, by 1925, was now working in the mines in Leadana and was underground along with many of the area men when the storm tore through.

"Dad said they had to climb up the zigzag ladders because the power had been knocked out to the mine elevators. They would stop about every 16-20 feet on the landings so they could rest. They were climbing so hard and so fast that they were wearing themselves out," Peachy said.

"Dad said he just kept thinking 'They're all gone,'" Peachy recalled her father telling her.

The miners had to travel by foot into town, much as Clara Brown and her parents had done, due to the enormous amount of debris left behind in the destruction of the mining area. Along the way the miners were stunned to discover boxcars and other freight cars having tumbled off the tracks of the rail line that wound its way past the mining village and through Annapolis on its north-south route through Missouri. Many miners noted that the boxcars were not only lying on their sides away from the tracks, but many appeared to have been spun several rotations away.

Peachy remembered what her father said he experienced when he arrived in town.

"He saw a young man on our property. He'd been walking around in the rubble and Dad asked him what he was doing. The young man said he had seen a heating stove on top of a bed, and he didn't know for sure if there was anybody there in what was left of the house, but he went in anyway and got it off the top of the bed before it started a fire. He told Dad, 'I hated to mess with anything,' likely because he didn't know whose property it was, but, he said, 'I went ahead and took it out.'"

Peachy's father found his wife sitting amidst the rubble covered in blood from minor head wounds and with the hole in her back, the baby Archie crying in her arms.

"Her face was all scratched up," Peachy observed. "Archie Ray would look up at her and cry, and she'd wipe the blood away from her face and he'd recognize her and stop crying 'til the blood would cover her face again. They kept doing this until Dad got there."

Peachy told of people who were crawling out from under wreckage all over town.

"No one said 'Are you okay?'" she said.

"They were all too dazed."

*　　　*　　　*

Sixteen-year-old Clara wandered through what was left of Annapolis, calling and searching for her fiancé, Carl Brown. She eventually found him as he reached the edge of the town along with other men from the mine. Carl was shaken but unscathed by the nightmarish storm that had demolished the little mining town. Together, Carl and Clara discovered what had happened to members of his family during the storm.

"His cousin, Osro Kelly, had his wife Nell and their two little girls, and they had gone into town that day in their old car and had just come back to their house. Nell and the little girls had gone on into the house, but Osro had decided to wait on the porch and see what was happening," Clara recalled.

"The little girls were okay. Nell was hurt real bad; she was laid up in the hospital for six months—they didn't think she would live, but she did. Osro was killed because he stayed out on the porch. Carl's mom took the two little girls in while Nell got better."

Osro Kelly was one of four killed in the little village of Annapolis, Missouri. Two others from the mine village, who had not been below ground, were victims, along with another niece of Carl Brown's.

The total dead, 4, numbered exactly one percent of the population of the 400 living in the village at the time. Twenty-five total were injured. Property losses totaled $400,000. Little Annapolis, seated in the hills of the Ozark Plateau range, suffered 90 percent overall destruction, meaning that every building but ten percent of them was either obliterated or damaged in some way, a phenomenal number of structures affected.

When the tornado left Annapolis on its heading of N 69 E degrees, it had slowed a bit from its entrance and was traveling at a forward moving speed of 67 miles per hour. Experts say winds were in estimation of close to 300 miles per hour within the vortex.

*　　　*　　　*

"I don't think we went back to school that year," Peachy reminisced. "Leadana lost the mining offices, doctors' offices, lots of housing buildings, a store…but it wasn't all necessarily rebuilt. The mine machinery was mangled."

The lead mine, which had been in operation from 1918, had not ceased operation at that point but had continued in a limited capacity until 1941, when it finally shut down, or, in Peachy's words, "it folded." Coal and other ores, however, continued to be mined in the immediate Annapolis vicinity.

July 3, 1999
leaving Annapolis

Something we noticed about K Highway between Redford and Annapolis—

There were homes along the highway, mostly few and far between. But whether poor shanty-type structures or rambling ranch-style brick houses, many bore a similarity. Within running distance of these homes were little grassy mounds, some built into the stone enclaves that bordered the homesteads, the stone being knuckles of the hills that were the edge of the Ozarks. Others had been deliberately formed by a backhoe in the middle of a pretty summer lawn. All of the little mounds had heavy wooden doors, hinged, with a pull handle of some sort that could be gripped easily and used even in the case of very strong wind.

In nineteen hundred ninety nine, and from years hence, the mighty storm that tore through the foothills of the Ozarks left an indelible mark on the memories of the residents and on many of their properties. It was not a mark left as a result of the crushing winds. It was the result of humanity's reaction to that wind—in the form of storm cellars that dotted the landscape as we cruised by.

The author's son, Jesse Wingard, posing at a community storm cellar outside of Annapolis, Missouri, in July 1999, in order to give perspective on the size of the underground structure; in the background elements of a park setting, showing the importance of such shelters to the residents of the community, which disaster struck on March 18, 1925.

Chapter Three

Biehle, Missouri

According to a chart developed by Stanley Changnon, Jr., and Richard G. Semonin in the book *Illinois Tornadoes*, after the tornado left the Annapolis area it struck an area called Cornwall that was somewhere between Annapolis and Biehle—but was no longer on any map. In the Cornwall area there were no deaths or injuries recorded, but damage was sustained in 1925 dollars in the amount of $1,000—so it was, it could be easily surmised, likely a rural farm area.

In and of itself, Biehle, Missouri, wasn't easy to find, but it was easier to find than Cornwall, which we decided to bypass in our search and go on ahead to the next stop of the storm as it carved its way into the record books. Biehle, pronounced Beel'-lee, was finally accessed at about 7 p.m. on Saturday evening July 3, 1999. Off State Highway 51, which headed north into Perryville in Perry County, there was an exit onto KK Road. KK went to B Road, appropriately enough, which went directly north into Biehle.

There wasn't much to see, maybe there never was. But there was a very large 'everything' store, Buchheit's, which was already closed for Independence Day weekend, and a couple of taverns (one, we discovered later, had been shut down recently). At the edge of town and across from the everything store was a stunning Catholic church. Up the way from the church was a well-laid-out cemetery.

Having at one time lived in Germany for many months, I could now discern a definite European influence in the plots of this cemetery. On this basis, I sat in my truck at the foot of the cemetery and took some time to gaze upon the neat rows of plot groupings, families buried together over many generations. Several of their graves were marked with large, flat stones over the whole of the grave. This was very European.

It was here I came upon the idea of searching the cemeteries for gravestones that were marked with the date of the killer tornado. The cemetery stretched back into a vast tree line in Biehle. The weather was so miserably hot on this date that it was insufferable to wander in and around the tombstones, and yet that was invariably what it would take in order to find one marker that was engraved with the date of death of March 18, 1925.

The grave marker of a Biechle family member killed by the Tri-State Tornado – Joseph Biechle.

For a half hour this was what I did until finally, I resigned myself to the fact that, of the many generations buried here, some dating back to the early 1800s, I was looking for a needle in a haystack once again— four deaths in a 1925 population of 100. I turned to walk back to the truck and noticed one small nosegay of faded pink rosebuds that had tumbled off one of the short, European-looking stones. Upon straightening the bouquet and sticking it back into the ground in front of the stone, I read the name Joseph Blechle engraved in the gray granite.

Directly below the name was the date of birth, June 6, 1883.

The date of death was March 18, 1925.

"It was so black, it was like pitch...with straw...and fence rails...just roiling"—Herman Blechle, age 9 on March 18, 1925

Herman Blechle was a fine old German gentleman. His English, which was quite good, was heavily accented with distinct German, the language spoken in his home from the time of his infancy. He was tall, sturdy for his 83 years, and he appeared the typical farmer, in his bib overalls and chambray work shirt. His face was weathered and lined, and tanned only where the sun could reach it from the shade of a garden hat or cap.

Herman Blechle, July 1999

Born in 1916, Herman was the third of four children; Edgar who in 1925 was 13, Gilbert, 11, and Adella, 7.

Their father, Joseph Blechle, came from Germany to settle in the area of southeastern Missouri. The area of that time had drawn so many from the old country; it was appealing in its similarities to the lower hills of eastern Europe, particularly, the Black Forest area after which cuckoo clocks and cakes were named. Several of German descent had already established decades-old homesteads in the country, growing and cultivating grapes that would be used for some of the finest wine to be produced in America. That remained a well-kept secret beyond the Missouri-Illinois area, but those regionally, and a growing number of collectors nation-wide, enjoyed the fruits of the vine to the full.

Joseph Blechle did not grow grapes, his son Herman said, but he did grow many other types of sustenance, a far cry from the 'farmer' of the twentieth century who, largely supported by taxpayer dollars in the form of "subsidies," worked thousands and perhaps tens of thousands of acres on an almost-industrial premise with mammoth machinery, only to sell the goods to someone to redistribute the fruits of his labor to major markets.

In 1925, what the Blechle family ate was mostly grown on their property, hunted, or fished in order to ensure their survival. Stores were for the procurement of what could not be raised or created on the farm.

And at the store in the village of Biehle was where Joseph Blechle was on the afternoon of March 18, 1925.

"He'd gone into town that day on foot," Herman said of his father. The walk was a short one; the Blechle homestead was located on a rise just outside the village and to the south; across Little Apple Creek, which ran northwest to southeast past the town. The property was situated on the right side of the road—a simple two-story structure with a cellar.

Many of Biehle's residents would build cellars in the weeks and years following the devastation.

*　　*　　*

In the shade of peach trees that grew in splendor next to outbuildings in the back of Herman Blechle's house, he explained what happened to his father Joseph and other members of his family on 'that day.'

"He'd been at the store," Herman said, referring to a little store/saloon owned by Edgar Stark at about the center of the village.

The village of Biehle used to have many such businesses over the years, but since that time, they were all wiped away with the removal of the railroad and the annexation of their village to the nearby city of Perryville. The only store left, by contrast, Buchheit's, was the large, kind of 'everything store.' This was the mainstay of Biehle as far as retailers; it was a farm store, had hardware, school supplies, clothing, a deli—you name it, Buchheit's had it.

But in 1925, "Dad needed to go to the store, probably got a beer while he was there. And we were in school; my two brothers and myself. My little sister was seven and she stayed home yet. There was a new frame school where the church is now," Herman said. The church, quite a regal structure, was St. Maurus Catholic Church, and was the highest point in Biehle.

"Mother was at home with my sister, and Dad was in town and ready to get home when he saw that it was getting dark in the southwest."

Farmers knew what that kind of weather meant, generally speaking—a dark, mysterious-looking cloud moving quickly from the southwest spelled nothing but trouble, but this had been such a mild, pre-spring day, Joseph Blechle didn't let his worries get the best of him. Instead, he headed off for home on foot, apparently feeling as though he would be able to beat the rain or wind or whatever the steel-gray cloud bore within it.

It was getting on to two in the afternoon, and the mailman was making deliveries in his horse-drawn cart at the time.

"The mailman," Herman said, "saw Dad walking and asked if he wouldn't like a ride up to the house instead."

Joseph Blechle accepted the ride. The air, Herman explained, began to feel stifling at about that time.

Others who were outside prior to the storm's arrival all concurred with this component of the tale.

Joseph Blechle sensed urgency and climbed up onto the bench seat next to the mail carrier. Together they rode on up the dirt road that crossed Little Apple Creek.

> *"It seemed like an hour. But in five or six minutes, the damage was done"—Julius Hotop, age 16 on March 18, 1925*

At a homestead located outside Biehle, there was an overpass for F Road above Interstate 55. There, Julius Hotop, his parents, and six of his bothers and sisters had resided and farmed. The house was situated on the side of the southbound lane of I-55, but as Julius pointed out, "There wasn't no Interstate or big highways like that back then."

Indeed, the resurfaced asphalt B Road that bisected Biehle probably seemed almost new to someone like Julius, who was raised on dirt roads and chat gravel when it was made available.

Julius, at age 90 in 1999, was as sharp as a tack. Another of German descent in the village of Biehle, he was a tall, well-built man who appeared at least twenty years younger than he was, in looks, mannerisms, speech and the twinkle in his eyes. His love for the descendants of his family, most of who lived nearby, kept that twinkle going. Having lost his wife in 1996, he lived alone in a tidy house next to his grandson Bill Hotop's tavern. Appropriately called 'Bill's Place,' it was the hangout for the villagers and served beer on tap, can, bottle, case and keg; they also had a decent menu and plenty of meeting space. It was here Julius Hotop traveled a couple of times a day to have meals. Many knew he was a survivor of the storm that swept through southeastern Missouri and that he knew those killed, injured, and left homeless by the force of nature; he knew all of them, even if it had been by mere passing acquaintance. The Hotop family were among the numbers who did not suffer catastrophic losses in the form of life, limb, or home, but the impact this particular storm made remained indelible even in the mind of one nearing a century of age.

"Yes, it was a hot day," Julius noted of March 18, 1925. "Earlier in the morning it wasn't bad, but toward the afternoon it got hot.

"I had a half-brother who lived in the river bottoms, and he wanted Dad to come stay with him and help clear out four acres of new ground, so that's what we were doing."

Julius Hotop

'We' consisted of Julius, 16, his father Joseph Hotop, brother Alphonse, 20, and Ervin, 18. They were used to the kind of work that was being done in the river bottoms.

"This was back when people raised everything they ate and ate everything they raised!" Julius noted with some amusement and some noticeable resignation to the path that the farm industry had taken in more recent decades.

The men had been working the field for several long hours and weren't paying attention to the sky, nor really the weather at all except to notice that it was growing unseasonably more warm.

"I'll bet it was a hundred degrees," Julius mused, speculating on the temperature. The likelihood was a good one; the warm front that preceded the storm and gave it the energy to move forward at locomotive speeds was settling quickly upon the Apple River bottoms and the temperature was rising.

Julius' brother, Alphonse, suddenly looked up from the heavy work and noticed the sky in the southwestern horizon bruising and growing closer by the minute.

"He said, 'Hey, we'd better go!'" Julius remembered Alphonse crying urgently.

The men were in the field with no team; they had arrived on foot and were working the land with only hand-tools.

"We were a quarter mile from the house so it wasn't too far to walk back,"

Julius said.

About half way, the men saw the approaching storm and how quickly it was bearing down on them.

"One of us said, 'We'd better run!!'" Julius recalled someone among their number saying, and, as the four men raced the storm toward the Hotop farmstead, "we got there just in time."

"It was so fast. It was real dark...Real dark"— *Barbara Buchheit, age 14 on March 18, 1925*

Barbara Unterreiner Buchheit was the oldest of four children of the Unterreiner family in Biehle. Also of German descent, she spoke with only a slight hint of German accent in her English, and then only with certain words and phonetic sounds affected. Greeted in German with *"Wie geht es Ihnen?"* (How are you?), she good-naturedly insisted "No, none of that. We speak English in this house."

The Unterreiner's modest homestead stood one-half mile from the current location of the little house in which Barbara resided, which was outside Biehle to the east, not far from the village of Old Appleton, Missouri. In the pre-spring days of March, 1925, the Unterreiner family had just built a new home five months before and had wintered in it just fine—however, there was no insurance on the dwelling.

Fourteen years old at the time, Barbara was in school with her brother, age 10.

"We'd already been in from recess for awhile and the teacher was the first one to notice that a terrible hard wind was coming," Barbara said. The teacher, she said, suggested to the students that it would be a good idea to bring them all up around her desk, no doubt to keep the schoolchildren away from the windows should the wind become strong.

"I thought to myself, I might as well stand up here (by the desk)," Barbara said, indicating the initial unconcern all were showing toward the first hint of the storm.

The unconcern, however, didn't last long; the forward moving speed of the storm was at that point about 67 miles per hour, and it was upon the little village of Biehle quickly.

"It was so fast," Barbara said, and after all the years, her voice was quite hushed as she recalled it. "It was real dark. Real dark."

*   *   *

At the creek which bordered the Blechle property, Biehle's mail carrier whisked his horse-drawn wagon across the bridge, stopped the carriage, and shouted to his passenger, Joseph Blechle, to run like hell up the dirt trail to his homestead before the monstrous black clouds that were approaching them from the southwest descended upon them.

Indeed, it appeared as though the clouds were merging with the ground just ahead of them—huge, rolling billows swept away the horizon and boiled toward them with demonic fury.

But Joseph Blechle just wasn't fast enough.

"The storm came down on him right then," Herman said, "left the mailman down there at the end of the drive, and took Dad up and into the trees, then threw him into the creek. At the same time it took Mom through the house, north to south, she said, and dropped her down into the creek bed with all kinds of trash."

Herman, from his vantage point at the schoolhouse, the highest point in Biehle, was able to get a clear view of the storm as it approached and of what it was carrying within it.

"It was so black," Herman said, with a voice that still reflected the wonder and terror of a nine-year-old boy staring at the face of such a violent weather event, "so black it was like pitch….and with straw…and rail fences in it…just roiling.

"The big steeple at the church, next to the school building, came down beside the school. And the top was driven six feet *into* the ground."

*   *   *

The amazing fury of the storm swept over and around the Hotop house—literally. Reports by engineers surveying the area following the storm in later months made mention of finding a 'double track' of the storm—two twisters that developed and made ground contact for 3.5 miles through and outside of Biehle; this was reportedly verified in the days, weeks and months after the storm by survey engineers, as noted in *Weatherwise* magazine. In modern-day weather terminology, this is termed "satellite vortices," tornadoes that spin off out of the original storm, sometimes maintaining, sometimes doubling back to become part of the original.

The Hotop home was about to become a victim of this satellite vortex phenomenon.

"It sounded like a roaring train almost," Julius said. "You had to see it to believe it.

"See, there was a steep hill down the side of the house, and the main storm, oh, it was about 300 feet wide or so, it came down across that big hill straight on, and swept past the house. But another tornado came around at an angle out of that big storm." With this said, Julius slapped a fly swatter on the kitchen table, indicating the farm house was being represented by the business end of the swatter. He drew with his right index finger a path of the smaller tornado traveling side by side with his left palm, which flattened out behind and to the left of the 'farm house.' The left palm became the larger twister, inching up toward the left side of the 'house'.

The right finger, portraying the smaller twister, snaked out away from the left hand/bigger storm, drew a rectangle to the right, then in front, and all the way around the farm house, and slammed back into the palm as they met head-on, ahead of the 'farm-house.'

SMACK!

"It swept around across the farm and was sucked up into the bigger storm. You had to see it to believe it!

"It seemed like an hour. But in five or six minutes, the damage was done. It tore fences out, and the livestock could go anywhere. It killed sheep. It took all the outbuildings, and tore the chimney off the house, and some roofing. It broke all the windows; and there was a corner of the house where barn timbers had run through," Julius said.

Yet, the house remained; many timbers from the demolished buildings had slammed flat into the house when the storm passed over and had created serious damage, but it left the house, and all the Hotops, intact.

*   *   *

The storm departed from the Hotop homestead, screaming toward the schoolhouse where Barbara Unterreiner stood with her schoolmates and teacher. It brought with it anything in the path that could be held aloft by winds in excess of 200 miles per hour—practically everything, including probably a few kitchen sinks.

Barbara, standing beside the teacher's desk, felt the school move "a little bit" as the storm swept over, through, and around them.

"It didn't tear it up," Barbara noted with a sense of amazement. "The school stood."

And as the storm moved on northeastward, the schoolchildren tumbled out of the building, looking for a way home, a familiar face, a sense of comfort to follow the tremendous power of the twister. Though the school stood, there was nothing around them that wasn't in disarray, if not utterly destroyed. It was as if some divine force had reached out of the heavens, protecting the house of learning and its precious inhabitants.

"We ran home," Barbara said of her and her ten-year-old brother. "It was two-tenths of a mile away."

When they arrived, they found their entire family standing in the yard. The new house was gone.

"The only thing left standing was the smokehouse. That storm took everything," Barbara said.

The miracle of the Unterreiner family was the fact that all of them had emerged unscathed.

"Mom was shaking, and she said that the last she saw of Dad and my little brother was that they were flying through the air," Barbara remembered. "Henry Wibbenmeyer, one of our neighbors, went looking for my brother and for his own kids."

Wibbenmeyer found the children, relocated by the winds several yards from their homes but unharmed. The Unterreiners' father was also with them, and also unhurt.

"When they came back, they found Mom and my baby sister Wilma. It was a strange thing, how they didn't get hurt. Mom was holding Wilma," Barbara said of her sister who was two years old at the time, "and when the wind came through, their chair fell over and a big homemade wheelchair, which was a good heavy one and sturdy, fell over on top of them. That wheelchair supported the ceiling that fell in on them. Then everything else, just about, was blown away. It was a miracle."

*   *   *

"When the neighbors found my mom," Herman Blechle said, "she was down in the creek bottom with all that trash, and she didn't know anything."

The statement 'she didn't know anything' was likely the nearest one Herman could find to use to describe the state of shock that his mother, Ida Blechle, was experiencing.

"She was all black and blue. Her dress was shredded. It looked like someone took a razor blade and ripped it all to pieces." Herman paused; it was difficult to discern whether he had beheld the next sight for himself or whether it was given to him as second-hand information; either way, its impact was obviously deep.

"The neighbors saw her walking in circles down there in the creek bottoms." Herman told it as if he were seeing his mother stunned and wandering. His eyes held a studied gaze that reflected the confusion that Ida Blechle inarguably experienced at the power of the tornado.

"She just didn't know anything."

Joseph Blechle was dead; his body lay not too far up the creek bottoms from where his wife was found wandering. Some have speculated in the years since if she hadn't perhaps come upon the body of her dead husband in her wandering and that was what might have sent her into shock—not that the powerful winds of the storm, which struck with a forward moving speed of around 67 miles per hour as it descended upon Biehle, wouldn't have been enough to send a sane woman into shock on their own.

"The house was gone," Herman said, "and everything in it was gone, and so was the roof off the barn."

"No one had ever conceived of anything like that (storm) before," Herman insisted.

His eyes had a faraway look that revealed more than Herman probably intended for them to do.

"They took Dad's body to a guy down in Biehle, and the only thing I really remember about it all is them sprinkling Dad with holy water after he was dead."

Herman didn't particularly remember the funeral, or the interment of his father, Joseph Blechle, in his final resting place in the front of the cemetery.

*　　　*　　　*

Aside from some dead sheep, life on the Hotop farm was preserved through the monster cyclones.

There were, however, some oddities, the first of several, noted by the Hotops.

"We were out hunting up the animals and while we were looking at all the damage, we were seeing ice driven into the ground, about this deep," Julius Hotop gestured a depth with his hands of about six to seven inches.

"They were just hunks of ice, not little round hail pieces that come down before a storm, these weren't nice round shapes." Julius Hotop's face took on a dark expression as he remembered. "They were about eight inches across in places, and they were in all odd shapes, big, jagged pieces of ice that were just buried in the ground. If one had hit any one of us, it would have killed us right then and there.

"The sun was out right quick after the black was gone," Julius reported. Several survivors later recalled that as the day wore on after the storm, the temperature dropped dramatically, and in some areas, it was rumored that snow actually fell. But in Biehle, this was not the case; the ice that pummeled the ground at the Hotop place through the duration of the storm was the coldest it seemed things were going to get.

"The temperature didn't drop that much after the storm. I know that because, that storm, it tore all the flues out of the house, and we didn't have any electric anything, so we weren't able to build a fire anywhere, but it wasn't cold.

"None of us stayed in the same place, though. I went to stay with my uncle Bill Schemel that night; the rest of them stayed in one place or another," Julius remembered.

*   *   *

Barbara Unterreiner Buchheit agreed with Herman Blechle's earlier analysis in that "Not too many storms go through here," she said. "We never saw anything like it before or since."

Barbara's future husband, Edwin Buchheit, was 15 years old at the time and lived three miles northeast of the path of the storm. Edwin died in 1978. His son, Richard, related the story his father remembered and shared with his family through the years.

"Dad said that he and grandpa were sitting on the front porch watching tree limbs and tin blowing through the sky....floating down to the ground. They were unaware of what was happening. They just knew that it was dark, and the wind was blowing, and very high in the sky there was trash, and it was coming down."

Richard pointed out the spot, thanks to his parents who etched the memory in the minds of their children, where the storm barreled through the countryside and left Biehle behind as it sped toward the Missouri state line with Illinois.

55

Destruction of Theo. Holschen's residence from the west outside Biehle, Missouri, following the March 18, 1925 storm. *Courtesy of the Missouri Digital Heritage Collection, Secretary of State's office.*

On the gravel road that lead to his mother's home east of Biehle, he stood at a high point and indicated, across Interstate 55, a barn and another outbuilding that withstood the brutal attack nature brought upon Biehle that March day in 1925. The two buildings, heavily weathered with time, had been shored up over the years; one contained large hay bales, the other was empty. They sat not too far from where the homestead that belonged to the Hotops once stood.

"Dad always said that this was right where the tornado crossed," Richard said. "There wasn't an Interstate here then, but he said there wasn't much of anything after that tornado went through, either."

Odd stories abound regarding any tornado, and this one was no exception. Along with the jagged, giant chunks of ice Julius Hotop and his family beheld, another survivor, Rudy Buchheit, owner of Buchheit's Store in 'downtown' Biehle, remembered another odd event. Rudy declined to talk about the storm or its effects, but his wife, Irma, did:

"The Buchheit's house was destroyed," Irma said, "but they had an oven/stove upstairs that fell through to the cellar instead of getting blown away.

"There was bread baking in that oven, and when they went back to the house to see what was left, they found it down there. The bread was done, and it was edible."

No doubt the Buchheits needed every edible thing they could find following the storm, as a large majority of their property and belongings was damaged or simply blown away. The storm struck little Biehle, Missouri, at exactly two in the afternoon on March 18, 1925. Official reports verified what Julius Hotop indicated, that upon approach into Biehle, two large funnels were seen within the storm instead of one. Slowing from its forward speed of 67 miles per hour to around 60, it took the lives of four of Biehle's 100 residents, injured 11, and caused $45,000 in damage.

From the point where the storm left Biehle, it crossed a 'rural area' of southwest Missouri, south of Longtown, north of Frohna, and west of Wittenberg, in so doing, ravaging a small community called Ridge.

Chapter Four

Ridge/rural area to Illinois-Missouri state line,

"After that I'd go outside and look at the sky. The clouds didn't have to be very big to scare me"—Oleida 'Lee' Rauscher, age 8 on March 18, 1925

Ridge, Missouri, was a residential community located about five miles north of the road that ran through Frohna and Altenburg. It bent toward the river, and sat in the vicinity of Wittenberg, a town which no longer existed but which was noted in several weather surveyor's reports as being alongside the path the tornado took as it crossed southeastern Missouri.

Ridge wasn't exactly a village in 1925, but was a viable community nonetheless, with three or four houses, a small country store and a church/school building. The charm of Ridge in the late days of the 20th century, while no longer necessarily in existence as even a viable community, was the fact that the Steffens Family Orchard grew and sold apples at a fruit stand near the intersection of Altenburg Road and C Road.

Another charming feature that originated in Ridge was Oleida "Lee" Rauscher, a warm and inviting lady of 83 in the year 2000 who had moved to Cape Girardeau in her later years. Her days in Ridge were full of family and a pleasant life. With four brothers, a sister, a half sister from her father's first marriage and both parents, Anna and Ludwig Gerler, it was a busy, happy family, most of who were in the Ridge school on March 18, 1925 when bad weather began to blow up from the southwest.

It had been an ordinary day, Lee Rauscher said. The walk to school was just a half-mile from her home, and her brothers Marvin, 11, and Erwald, 9, had accompanied her. Lee's half-sister Elda, 13, had been attending school in Altenburg for confirmation. The building in which the Gerlers attended in the company of other students for a total of 22 others was also used as a church for the Lutheran community, most of which originated in Germany or Austria, as Lee's family had. It was a strong Lutheran community, and Ridge was a parochial school taught by a young lady who stayed with a nearby family.

It was this young lady, the teacher, who noticed something ominous in the air shortly after 2 p.m. when clouds began to creep up along the western horizon. The fine spring breeze that had accompanied the Gerler children and others to school and had warmed their recesses and lunch hour had begun to pick up and become insistent. Lee remembered the schoolroom becoming so dark that the pupils could hardly see to continue their early afternoon studies.

"I remember everybody whispering, 'It's a storm coming up,'" Lee said, "and everybody looking out the windows."

The sky was darkening quickly and more heavily than any storm any of them had ever before seen in their young lives.

But the teacher was determined to maintain control, if not over the elements which threatened them, then at least over her charges within the classroom.

"She was pretty well aware of what could possibly happen," Lee said. "And so when the door to the room blew open, she really calmly asked one of the older girls who sat near the back to go close it."

The door was in the back of the schoolroom. The girl walked to the back of the class, closed the door and went back to her seat. Just as she sat down, another violent gust of wind caught the door and it burst back open.

"Please go and close the door again," the teacher told the girl sitting in the back.

Again the child did as instructed. But again, the door was forced open by the torturous winds, and now the students could see large tree limbs blowing past the schoolroom window and to the ground, only to be picked up and carried away by the amazing power of the storm as it approached. With the door standing wide open and the heavy gusts coming through, schoolwork and papers began scattering about the room. It was at that point the teacher instructed every student to move to the back of the room and hold the door shut with all their might.

> *"When I was growing up, if anybody talked about it, I'd walk away"—Cordia Nebel, age 6 on March 18, 1925*

Cordia Nebel

Cordia Stueve Nebel, one of the youngest children in the school at age six, recalled in May of 2000 more vividly than anything on that March 1925 day the fact that their teacher must have known what she was doing when she moved the entire student body to the door to keep it closed.

"I remember so much trying to hold the door shut," she said.

She was convinced, she said, that if they'd left the door open, the tornado would have come into the school and killed them all.

At this point, tree limbs were crashing against the windowpanes and breaking out the glass, and all the children could see the storm barreling at them from the destruction it had so recently left in Biehle, when it had divided into two twisters and had devoured most of the tiny village.

The children had no real way of knowing, but they could at least see there was something unusual and distinctly ominous about this monstrous storm, which came at them like a mass of thick fog rolling over the landscape. Rain was falling in a torrent, but the heavy and steady bursts of wind were driving it sideways and the appearance of it all was like a very black coal smoke, according to at least one witness.

Cordia was aware of the damage occurring all around her, but was nevertheless taken extremely aback when the very school building began to shake and rock with the force of the winds.

*   *   *

Lee's recollection was very clear. When the shaking of the building began, she fell between two of the desks, which on Sundays served as pews.

"There were drop-leaf tables in the back of each desk," she said. "The last thing I saw while I was standing was the wind blowing the windows out and just everything coming in through the windows.

"Then I was aware of something heavy hitting me.

"It was not so bad, the pain at that time," Lee admitted, although her voice was tinged with the reality that pain was to come in abundance soon thereafter. "But at that point, I remember nothing except that it was raining like everything. By the time the wind picked up the school, it was just raining like everything."

Lee did not exaggerate when she said the wind picked up the school, for that was exactly what happened. The entire Ridge Parochial School and Church was picked up by the force of the monster tornado and carried, its precious cargo of 22 students and a teacher who thought she could keep the wind at bay by having the students hold shut the door, from its foundation and hurled through the air like a child tossing a toy across the hillside of the area of Ridge.

*　　*　　*

"We went up over the trees, over the road, and onto the hillside," Cordia said. "When the schoolhouse hit the hillside, it scattered us all."

The school landed several yards east of its original location. According to reports, there was nothing left of the framing; the building was totally in pieces including the floor and the floor joists that had come apart from the structure and came flying after the students as they landed, littered about the hillside.

And they landed to Cordia Stueve's great dismay—and injury came and violated their little bodies.

"The floor from the school fell right on me," she said. "Oh, I was hurt alright, but another girl, she was scalped by that falling floor. But me, well…"

Cordia tended to downplay her injuries inflicted by the flooring, but they were visible even in the decades following the storm, and they were nothing she could deny.

"The floor just scraped up my left leg, took the skin right off the bottom part of my leg, and broke the bone between my hip and knee. And I guess I had a broken collarbone, broken ribs—I don't know if I was broken up inside, but it was pretty bad all the same."

*　　*　　*

Amid those same pieces of Ridge schoolhouse was Lee, who remembered lying within the rubble of the building.

"The rain was coming down in sheets. I noticed one of the older girls was walking by and didn't see me, but the Lord made me pick up my hand and I grabbed her dress."

The girl pulled Lee from the destruction and they held hands, wandering around for a little while.

"She wasn't sure where she was," Lee said, uncertain as to whether the girl realized the entire building had been relocated to the adjoining hillside. "Then she got her bearings straight, and we went to a house nearby."

While Lee recalled the home she and her savior went to was occupied by Edmund Weber, who boarded the schoolteacher in his home through the school year, Cordia insisted the house the girls retreated to was one in which her own grandmother resided, having remarried Carl Leinbach.

"It was a small house, with an upstairs," Cordia said. "I think Leida remembers it because the house was so damaged. Edmund Weber's house was, too, but my grandma's house was in bad shape. They took a lot of us there, and laid us out in what was left of it, but the kitchen was totally gone. There were just two rooms left downstairs, and the two upstairs."

"By the time I got there, I was in pain," Lee remembered. "That girl, and I think it was Elda Engert, was helping me up a little step to get into the house, and I almost fell into the cellar. I got in there and leaned over something to ease up my back a little."

Lee remained uncertain about what it was that landed on her so heavily. But in later years, her trips to the chiropractor became frequent, and provided, while a little relief, not a complete restoration of function from the injury her back sustained.

Her friend Cordia, however, was not so fortunate.

By the time to skies began to clear and the torrential rains began to dissipate, the community of Ridge was beginning to realize something dreadful had happened to their children in the schoolhouse. Many rushed from their homes and came to see if their children were all right…or if their worst fears had come to pass.

"They laid me aside," Cordia recalled, "and eventually my dad came riding up in a wagon. There was a kind of a daybed in it, you know, like a lounge chair, that they put me in. But I couldn't stand riding in the wagon; it hurt too bad. So they took the bed out of the wagon, and some people who had come along carried that bed with me on it all the way back to my house."

The house was located two miles away from the original location of the school. Lee remembered seeing her friend Cordia go by, being carried by men of the area on a "cot," as she recalled; they passed right by her where she stood, having been brutalized by the storm.

"I could see her out there," Lee said, "and it looked like she was really hurt bad, so much blood. It looked to me like she was spitting blood, maybe hurt on the inside."

Cordia didn't remember spitting blood, although her injuries, including broken ribs, could have facilitated just that. Instead, she does remember her dear aunt, Mary Stueve, walking right beside her every step of the way, and offering her wine to drink, as often as she wanted it, to ease the pain.

"We made wine back then, and a lot of folks said it was the best wine around," Cordia related. "My aunt Mary was so good, staying right there and talking to me to keep me calm and take my mind off the pain, and giving me wine."

They finally arrived at Cordia's home.

"They took her there because back then, people didn't just take you to the hospital, especially if they didn't think you were going to make it," Lee said. "I think they thought it was better to die at home."

"But I didn't die! —obviously!" Cordia laughed. "They kept me there for a few days, and they tried to treat me. There was a doctor up there, but there were so many hurt…"

Cordia related that while many were injured, no one was killed in the Ridge area as a result of the massive twister. But the extent of the injuries kept the doctor very busy.

"My folks put boards on each side of my broken leg to hold it, and that's how I got scarred," she said. "I don't remember crying, and I don't remember being in a lot of pain."

What she did remember was after a short time, her family took her to St. Francis Hospital in Cape Girardeau, where a doctor Fruth or Furth (the exact spelling and pronunciation of the name escaped her) "brought me back to where I am today. They took good care of me in the hospital, but I don't remember much of that, either."

The effect on Cordia's leg was profound. In the decades to come, when any doctor addressed the injury and wanted to know how it was sustained, Cordia would explain, and on many occasions the staff would call her "The Miracle Lady" thanks to her survival. Still, her left leg remained a reminder of that dark day.

"A lot of doctors have looked at it to see if there was any plastic surgery that could be done on it. But they'd say, 'It's such a big job, there isn't anywhere we could get that much skin from' to do grafts. So my left leg is thin, not nice and round like my right leg. But I've made it all these years."

The emotional scars enveloped both girls in the days, months and years to come.

Lee recalled, "After that, I would go outside and look at the sky before I'd go *anywhere*. If there were any clouds, and they wouldn't have to be very big, I wouldn't go.

"But my mom used psychology on me, although I'm sure she never took it, and she let me stay home and let me see for myself that there wasn't anything to be afraid of. I had to make myself get over the fear. I'd say, 'It's not going to be a tornado.'"

Cordia had a similar response.

"I always cried every time it would thunder, and Mom or Aunt Mary would hold me until I stopped crying."

But like Lee, Cordia, within a years' time, came to work through her terror of storms.

"One day, I was about seven or eight, I was by myself, and no one was around to comfort me when it thundered. So I made up my mind and said, 'I'm not gonna cry no more. I'm too big for that.' I didn't see nobody else crying, and they were in that tornado, too. So I got over it. But when I was growing up, if anybody started to talk about it, I'd walk away."

The twister swept through Ridge and injured many, some severely, but cost no one their lives. Through the 'rural area' outside Biehle and before the Mississippi River, however, in the vicinities of Altenburg, Frohna and Wittenberg, it killed a total of four, injured 25, and caused $113,000 in damages before it approached and crossed the mighty Mississippi.

It had left at least two little girls with a healthy fear of the unknown that weather can bring. But the Missouri-born twister was about to tread heavily upon yet another state as the date March 18, 1925, moved into the history books as one of the deadliest weather days on record.

Illinois

Chapter Five

Gorham, Illinois

After leaving Biehle, my children and I stopped for the night in nearby Perryville, a lovely town based on industry which sites were laid out in neat locations on the edges of town, thriving on the I-55 traffic stopping for a meal or a night of rest. The drive out of town toward the nearest Mississippi bridge was hilly and scenic until just before Missouri Highway 51 gave way to the bridge. Then it flattened out, and became the river bottomlands, productive farm acreage that was gleaming green in the Independence Day sun.

Crossings of the Mississippi were infrequent between Missouri and Illinois in the southern part where we were traversing. The nearest bridge that took us to where we were going was in Chester, where some of Illinois' most vile criminals resided in the Menard State Penitentiary.

It was the town known for Popeye the Sailor Man (the creator thereof having come from the town) and river barges. It was also one of the many towns off Illinois Highway 3 that lay near 'The Bluffs.'

That drive was breathtaking, no matter how many times one made it. The Bluffs were walls of rock carved millennia ago by the rushing, perpetual waters of the Mississippi, high 'riverbanks' that tapered off flat at the initial base. They were wide enough at the base to have enabled the state to create a roadway—Illinois Highway 3. Then, from the roadway, more tapering off occurred, and sloped quickly down to the immediate Mississippi riverbanks.

Highway 3 was a two-lane upon which drivers traveled at a high rate of speed. There were several locations along The Bluffs where the rocks had come loose and the pieces had drifted down until they made viable rockslides, and this was indicated by highway signs posted in the danger zones. Drivers would zip around curves created by the river until they were out of the way of the potentially falling rocks. A modest little village, appropriately named Rockwood and a former timber market, was the only real civilization on Highway 3. Little Rockwood was known for Footprint Rock, a presumed Native American creation in which footprints of differing sizes, as well as an imprint of a hand, were chiseled perfectly into the rock in the bluffs at a point

Welcome sign to Gorham, Illinois.

just outside the city. Many decades after their discovery, several of the footprints, and the rock surrounding them, were said to have been hewn out of Footprint Rock and taken by vandals. Whatever the story, it was an interesting piece of history from an area that held a population in the range of only about fifty residents all told at the turn of the twenty-first century.

Between Chester and the turn on to Illinois Highway 149, which took the driver away from the western side of the state, Highway 3 continued south beyond that point. Past that turn we found what looked like low mountains or tall hills to the east; these were technically part of the Shawnee National Forest that covered the lower portion of the state of Illinois. These were areas of the state that were unglaciated when the Ice Age glaciers receded, having created the prairie, pushing up rock-monuments just to let mankind know they had been there. Little Grand Canyon was located twelve miles inside the ridge of hills and was a beautiful, though somewhat dangerous, place to hike. A winery and campgrounds were also found in the immediate vicinity. Logan Hollow Fish Farm was the last landmark to look for on the left, over three miles past the Hwy 149 turn off Illinois 3. It was quite large; several buildings greeted visitors off the highway, but spread out beyond and around them were symmetrical, small ponds, providing homes for different kinds of fish. The summer we were there, most ponds were full of water; several had been drained in the hot sun and were being carefully plowed by small tractors. This was to keep the danger of contaminants down and the balance within the pond as clean and natural as possible when time came for the next fish stocking.

On the opposite side of the fish farm was the first set of bluffs to lie on the west side of the highway. This was also where the turnoff toward Gorham was situated.

Rounding the bluff was not something to be done quickly, not just for safety's sake but so it could be enjoyed to the fullest. The trees in the summer were lush and covered a good portion of the sides of the bluff with thick canopy. In several locations, mimosa trees, with their delightful, fluffy pink blooms, jutted forth from what appeared to be a wall of solid stone.

The stone was everywhere on the bluffs, in sections, rising straight and vertical. The ground immediately below was impossibly flat and covered with the fruits of the farmer's labor as it had been for well over a hundred years. It was as if the bluffs outside Gorham chose to rise one morning with the dawn, breaking forth from the surrounding flat ground that bordered the Mississippi with a terrestrial growl a shout and finally a roar. They rose high into the sky and left the flat ground behind to forever gaze up in awe at the incredible rock formations that formed a barrier from the north winds that attempted to sweep down upon the few houses that were settled just south of the rocks.

Beyond those homes lay Gorham. A village of about 400 at the end of the twentieth century, it had seen its share of hardship and tragedy in the past century. A storm, arriving from approximately the same location as the Missouri cyclone, crossed the Mississippi and bore down upon the village in 1957, wreaking havoc but not nearly on the scale as the storm more than a quarter of a century earlier.

In 1993, a flood of almost Biblical proportions struck the Mississippi River valley and pressed at a flood level of 49 feet against a 50-foot levee. Residents claimed, and rightly so, that the flood of '93 did the town in. Many people moved out and never came back. There was too much risk to the village, which, if the dam had burst, would have been twelve feet under water. Flood insurance being expensive, the working poor in Gorham could not afford another major rain to wash away everything they had worked for all their lives.

So they packed up and left, deserting the village for higher points and more secure employment venues.

In 1999, tragedy again struck the village as Angel Maturino-Reséndiz, the "Railroad Killer," made a brief stop along one of Gorham's many rail lines and reportedly took the lives of one of the village's elderly residents, George Morber Sr., 80, and his daughter Carolyn Frederick, 52, in their homes only yards from the tracks June 15, 1999. The search for the "Railroad Killer" ended mere weeks after the attack on Gorham, with Reséndiz turning himself in to authorities in Texas. But the mark he made was a permanent one on the remainder of the little town that had fought (and sometimes lost, sorrowfully) against incredible adversity.

There were few viable businesses left in the town that once had a population between 800 and 900 just before the turn of the 21st century. There was a downtown market that was more closed than it was open; there was a very small bank within the same building that housed the post office; there was a United Mine Workers of America and a Village Hall, and literally no indication that there was ever a serious railroad business except for the many tracks and switching points that sent trains on their way out of Gorham to Chester, Murphysboro, Cairo, and places beyond. But there was at one time an incredible trade in the railroad industry going on in Gorham in 1925. And it was railroad workers and the families of those who had settled in Gorham by 1925 who were hardest hit of all by the raging March 18 storm.

> *"People were in a daze. They couldn't believe what they were seeing and what they were living"—Rosetta Adams, age 9 on March 18, 1925*

"Gorham used to be such a pretty place," remembered Rosetta Casey Adams of her hometown in the early 1920s. "There was a lot of pride here. People kept the town really clean."

Of course, the incentive was there, as it appeared Gorham was on the verge of becoming a major force in the burgeoning westward move and the gateways that linked the communities on the east and west sides of the Mississippi.

"There was every kind of business here," Rosetta said. "The railroad of course was the center business. It was the Illinois Central railroad and the

Missouri Pacific. We had a nice depot and there were passenger trains every day that ran in to Murphysboro, and up to Bush and Hurst.

"Then we had Schimpf's Meat Market, the General Store, and it had dry goods and yard goods. We had an ice cream parlor, a barbershop and two restaurants. The bank and the post office were in the same location they are now, but up the street Slim Killion had a movie house in the building that's a church now, and the movies for the kids were upstairs and they had pool tables downstairs. And then they had two garages here in town, and a filling station, and Crain's Lumberyard at the end of town."

Rosetta's father, Ed Casey, was a railroader, "a section hand," Rosetta said.

"He had the roughest job. He was with the men who got down in the muck and worked tamping down the railroad ties. In fact, he lost an eye tamping ties into the creosote. Creosote flew up into his eye and put it out.

"My mom was an ol' 'root woman,'" Rosetta Adams said of her mother Edna Casey. "She would treat people that doctors couldn't or wouldn't help. One time, when I was still pretty little, she was called to help this man who had gotten trench foot in World War I. He had tried everything, had been to the Veteran's Hospital, and nothing could be done about it.

"So Mom took me and we got out in the bluffs and dug up some roots, some blackroot and some ginseng and a couple of other herbs, and she ground up everything and mixed it together in portions. Then she got some pure butter and made a salve, and took it to this guy and had him put it on his foot.

"And he got hold of her later and told her, 'There hasn't been anybody or anything that could help me until you came along.' And he thanked her."

Edna Casey was also a very successful midwife. "She delivered almost every kid in town," Rosetta said. "She would sometimes walk for miles to deliver. She was good, never lost a mother. Delivered a few babies stillborn, but never lost a one."

But Edna was "the most rambunctious person I ever knew," her daughter claimed. In the spring of 1925, Edna Casey was a young 31, and full of life and vitality and felt the need to be constantly on the move—literally.

"Mom picked up and moved every six months. She couldn't sit still," Rosetta revealed.

The Caseys owned a home in town, but Edna was not satisfied to live in one place for any extended length of time. She wanted to be on the move. It was never very far away, but she was going, going, all the time.

"I understand why now," Rosetta said, but didn't reveal the exact reason. "I know that she wasn't very happy. That caused her to be on the move, always going."

It was such a move, prompted by Edna Casey, on a warm, pre-spring day that Rosetta remembered was what saved the lives of the Casey family.

> *"Grandma was found after the storm went by.*
> *They put her on a railcar and handrailed her down*
> *to the depot"—Lucille Thurman, age 7 on March*
> *18, 1925*

Lucille Thurman, 81 in December of 1999, remembered fondly the little town of Gorham that had a big future with the railroad back when she was a girl.

"Grandpa (C. C. Crain) had Crain's Lumberyard," Lucille said, "and at the corner he also had a filling station."

Business was good for Crain.

"The Illinois Central Railroad, which ran north to south, had a radio tower on the way out of town, and the Missouri Pacific, which came up out of the south and went north at night, and the Cottonbelt Railroad, which ran east to west, kept everybody busy. The depots had three crews a day, and they used the restaurants, the rooming houses, the grocery stores."

The crews would stop in Gorham, eat, and then take in coal and water on board the trains in order to fuel and cool the engines. By any estimation, it was an important stop in Southern Illinois for many railroad companies.

"And there was a passenger train, the *Billy Brian*, that went from St. Louis to Benton," Lucille said. "It ran the track around the bluff."

Therefore, life was good for the Crain family and others like them.

Lucille even remembered, "Grandpa, down at the lumberyard, had the first Dynamo lights." Prior to that time, most used kerosene lamps. "He had the Dynamo generator," Lucille said, "so we had the first electricity in Gorham."

*   *   *

Rosetta Adams spoke fondly of her family who were leaving out of Gorham on the afternoon of March 18, 1925, as per her mother's request that they relocate.

"Dad just put up with it," Rosetta pointed out, reflecting the resignation Ed Casey no doubt felt toward his young wife's proclivity for running. "He had hitched up the team to a wagon that he had loaded up all our stuff in. It wasn't a very far move; we were just going to the edge of town, up closer to the bluff, but we had straw packed in the back of the wagon so that it wouldn't be too cool for me and my brother and sister," Rosetta said, even though earlier in the day, late morning, she had heard her father state, "Well, it's pretty hot today; hope it don't come up a storm.'"

John Casey was 7, and little sister, Mildred, was just 3. Big sis Rosetta was a mere nine years old at the time, but she remembered pulling away from their little home in the middle of Gorham. She particularly remembered riding the short hardpack trail north out of town when she and her brother and sister noticed the swollen, bruised cloud approaching from the west across the river.

Edna Casey noted it too. "She said, 'Ed, you'd better head the team up!' to Dad," Rosetta said. She continued that her father switched the team and they headed out a little faster.

"This was when we saw it, right before we rounded the corner at the bluff.

"It looked like a solid sheet. You could see white steam coming in front of it," Rosetta said as she indicated with her hands how flat and wide the wall was as it made its way across the river. Her description left the impression of how the walls of water might have appeared in Biblical times when God helped Moses part the Red Sea.

"The storm came across where the river bluff down there looks like it goes down into the river. It was so big, so wide," Rosetta said with clarity of memory.

Knowing what she did that day about the storm's development and how it rumbled through, up, and down the foothills of the Ozark plateau, she offered a comment based on what she saw at age nine and what she knew in 1999 at age 83 to be the case.

"It had been working through all those hills," she said, "but when it hit open country, it broadened out and hit full force."

Indeed, if it had not already up to this time, the storm had now taken on deadly and phenomenal proportions. Its forward moving speed was sustained at about 60 miles per hour, approximately the rate it was moving when it left Biehle, but in slowing, it began to build wind and vortex velocity, and the winds pushing ahead and within the vortex were estimated at a grinding 300 miles per hour or more. Up to now the storm had been nothing but blind fury. Now it evolved into pure focused destruction….and that focus was on Gorham, Illinois, population approximately 800.

Reports from the time indicated that there was a section on the west side of Gorham, up from the riverbanks, that had the very grass peeled from the ground, so tearing were the winds.

*   *   *

The storm resembled a wall of amorphous clouds, but were so strikingly distinct that they formed a rumbling wall that merely boiled flat vertically as it swept into the plain upon which Gorham sat, unwittingly awaiting its turn to be devoured by the beast the storm had become. Very little thunder was reported; only the heavy rumbling as the cloud, at this point about a mile in width, exploded into the floodplain.

"It didn't look like a tornado," explained Lucille Thurman.

Indeed, this tornado was apparently an enormous vortex very close to the ground and shaped like an inverted, truncated cone. It drove the river water ahead of it in a hissing, vertical rain, hence the appearance of 'steam' Rosetta Adams had noted upon the storm's arrival.

"We stood on the back porch watching the cloud," Lucille said.

Their home was located just around the block from the house the Casey family had just fortuitously evacuated.

"My parents didn't really have any idea what they were seeing," Lucille recalled. "There was no screaming or hollering. We just kind of huddled together under a table in the house."

*   *   *

Ed Casey whipped the team of horses pulling his family and all their worldly belongings. The horses broke into a wild run, finishing the distance down the road outside of Gorham and completing the curve around the bluffs. The Casey's new home was situated where the overpass for Highway 3 would be located several years into the future. Although there was no overpass at the time, the bluff stood like a waiting sentinel.

Fortune was smiling upon the Caseys on March 18, 1925.

"That bluff is a buffer," Rosetta insisted. "It turns storms."

The bluff was so tall and so thick, Rosetta and other villagers claimed, storms hit it and either slow down or reduce their intensity upon meeting its equal. The outcropping of rock, they believed, also served to diminish the severity of storms for homes located within its crags below. While the new Casey house sat to the north and west of the mighty rocks instead of right below it, believers felt the effect was the same. The powerful storm traveled the distance over the bluff outside of Gorham and went on its way, leaving the family of five behind—and unharmed.

*   *   *

"It was over pretty quick," Lucille Thurman said.

That remark was understandable. Moving forward at 60 miles per hour, the storm, wide and broad as it was, swept over the village of Gorham and moved out at 2:26 p.m. Given the conditions and speeds at which an average automobile travels, and the fact that drivers must go north on Highway 3 then south into Perryville in order to reach Biehle, which takes easily an hour, it was still mind-boggling that it took *26 minutes* for this storm to leave Biehle and reach Gorham.

Grandpa C.C. Crain was in the lumberyard when the storm struck, Grandma was nearby at the Crain home.

"Grandma (Jane Crain) was blown out of the house, by the bank, and onto the railroad tracks," Lucille recalled. Her voice heavy, her tone serious, Lucille said: "Somehow in all that, a two-by-four was blown through her side."

Lucille's sister, Jane (Crain) Feller, living in Fairfield, Illinois, in 1999, was a baby at the time of the storm. But the incident of what happened to her grandfather remained in vivid detail in her mind. As she grew up, she continued to live with the results of that incident.

"Grandpa was in the lumberyard and when that tornado came through, a bundle of shingles hit him in the face and head. A lot of that grit on the shingles got embedded into his face," Jane said. "It didn't kill him. But it tore his face all up. Disfigured him."

"They took him later to a cancer institute in Savannah, Missouri," Lucille recalled. "They did a little surgery at a time and then they had to take a rib and skin and make a jawbone for him. But it never did heal right. It wasn't an open wound, it was an open *heal*.

"I was a kid then, but I was scared of him; he lived with us for awhile. And then cancer set in. It looked so horrible."

Lucille's and Jane's grandfather died in 1952 of cancer. Their grandma, however, did not get the chance to bid her husband farewell.

"Grandma was found after the storm went by," Lucille continued. "They put her on a rail car and handrailed her down to the depot."

This was the passenger depot at which the *Billy Brian* was available to take folks down to Cairo, at that time in history a bustling town in deep Southern Illinois, at the convergence of the Mississippi and Ohio rivers. Those in attendance saw to it that Grandma Crain was made as comfortable as possible, considering that she was on a small, open handrail car with an old piece of timber, over a foot in length, jutting through her left side. Each end was so jagged it was difficult to tell where it had entered and pierced her—from the front or from the back.

Lucille Thurman remembered it with a child's simplicity: "They tried to help her, but she died en route to the hospital in Cairo."

*   *   *

"It was a good thing we were moving that day," said Rosetta in retrospect. "I wasn't in school because of that. Good thing I wasn't! The school was a bad place to be.

"Grade school and high school all went to classes in the same building. The top story of the Gorham school blew off; there were a lot of kids injured."

One in particular came first to Rosetta's mind, as the death of someone she knew resulted from a most horrific accident.

"It was the principal's daughter," Rosetta explained. "The school bell fell on her and cut her right in two."

A hallway, Rosetta pointed out, ran the entire length of the school on both floors.

"When the storm blew up, something heavy," she was unsure exactly what, "fell through the upper hall and made a big hole to the hall down below. And our janitor, Charlie Sumner, was down there right under the hole, and as kids took off running down that hall trying to get out of the school, they'd fall through the hole 'cause they didn't or couldn't see it, they were panicked.

"So Charlie caught 'em, one by one, as they fell through this hole!" Rosetta laughed heartily. "He caught all but one of the kids as they fell through."

The unfortunate child was Walter McMinnis—but he wasn't hurt.

"We always said 'you couldn't kill them McMinnis kids!'" Rosetta roared with laughter.

"Mom and Dad unloaded some of the stuff, and took us out to the new place and came back into town," said Rosetta.

An aunt came to stay with the children while their parents were away.

Their parents came back after having been in town all that grueling night. Even though Rosetta herself did not witness it, what her parents saw and what they told would forever haunt their eldest daughter, who was barely old enough at the time to realize what she was hearing when it finally came out.

With her mother's skills in nursing, she had been put to work instantly upon arriving in Gorham.

"Mom helped bandage up as many as she could," Rosetta said.

Her father, however, was given more grim tasks: that of loading the injured onto flatcars to send them to places where they could be helped medically, outside of the disaster zone; and finally of loading the dead onto more flatcars to be taken to Murphysboro to a morgue.

"When Mom first came back, she was telling her sister about it and she was shaking and crying," Rosetta remembered. "Dad buried it inside him."

Apparently, the sights and sounds her parents beheld were more than Ed and Edna Casey could handle. Even beyond the initial hours after the storm, when the dead had been shipped away and the injured taken to facilities for assistance, those left behind were facing wreckage beyond comprehension, and the Caseys were overcome by the enormity of it all.

They finally took their children back into Gorham.

"People were in a daze afterward. They couldn't believe what they were seeing and what they were living," Rosetta said.

Cousins of Rosetta's were but two of those she remembered suffering severe injuries at the hands of the wicked storm. Roy and Fern Casey were blown out of their house and into the woods directly north of Gorham.

"It cut Roy's face open, from his right ear to his mouth," Rosetta said. "And Fern's skull was showing. There was a scar over her forehead and back onto her scalp. We don't know what exactly did that to her. Being blown through the woods, it could've been anything."

Rosetta's uncle, Herman 'Dode' B. Casey, was out on the railroad on the south end of town, and, as Rosetta told it, "He knew by the way the storm went where to look for those kids, which way they had been blown, and sure enough, someone heard them crying from that direction."

The children's mother, Bert Casey, did not fare with such fortune. Bert died in the storm.

"We never did know what actually killed her," Rosetta explained with a sorrowful note; she was partial to her aunt Bert.

"But you know, the storm itself could kill you."

*   *   *

Lucille Thurman had little to add at this point. The storm swept its way on its direct N 69 E heading, taking itself away from Gorham at 60 miles per hour.

"The weather and the temperature were bearable after that," she related, "but it took a long time for things to get back to normal—it took a good year, at least.

"I didn't have nightmares so much; I wasn't hurt. But a lot of the survivors did need to have medical help," Lucille intimated events, revealing the hideous proportions by which the storm left its mark by comparison to the mere minutes it hovered over Gorham.

The storm was leaving the western edge of the second state where it had created unparalleled devastation. Gorham, home to over 800 people, became known as the first town to achieve the apocalyptic epitaph of 100 percent destruction, regarding the amount of buildings or homes destroyed or damaged; property losses amounted to $150,000.

Thirty-seven lay dead, and 170 were injured; a staggering number of dead or injured in one town by contrast to the 17 killed and 63 injured in one state, Missouri. The number of dead and injured in Gorham amounted to 41 percent of the population of the village.

That, however, was not going to be a record set for the day of March 18, 1925. In less than ten minutes, or roughly the time it takes to drive a speedy auto over the hills and into the city limits of Murphysboro if leaving from Highway 3, the monstrous storm was about to set new records of devastation that remain unbroken in Southern Illinois.

Chapter Six

Murphysboro, Illinois

"And Courage to Never Submit or Yield"—
motto, Murphysboro Class of 1925

On the drive out of Gorham in the stifling heat of the Fourth of July, we finally backtracked across country we had already visited.

We took Highway 149 off Highway 3. Highway 149 ran a slanted east to west from the edge of the state and into a line of hills that encircled Murphysboro along the west.

The hills were low enough to not show up prominently on any basic road map other than to be indicated by the shaded green of the Shawnee National Forest. Kinkaid Lake, Turkey Bayou, and the Lake Murphysboro State Park were here in what some fans of this forest lovingly called 'The Woods.'

The hills here were a lower version of the broad and rolling low mountains that were marked characteristics of the Shawnee National Forest, which covers a vast expanse of the southern fourth of the state of Illinois. They represent the 'unglaciated' portion of the state.

During the last Ice Age, the huge glacier that spread its icy fingers down through the central United States, left the flatlands and plains beneath it, and carved the beginnings of the mighty Mississippi also pushed great mounds of rock and earth skyward in this part of the state. The hills part of the state was considered "unglaciated;" that is, untouched by the glacier, but the effects can be readily seen. It was as if Mother Nature dug her fingers in and left lakes resting lazily between the hills in the imprints, and when she retreated, the hills reflected off the stunning spots of water, much to the delight of campers and fishermen from all over.

These hills with their gentle rises didn't cause as much automobile grief and consternation as did the ones that comprised the Ozarks; but there were a couple of points where any automatic transmission vehicle was going to feel a strain as it pushed itself up and up. Finally, atop one of the highest points outside Murphysboro, the transmission would get a break.

Notably, this very same hill, upon ascension going the opposite way, provided a breathtaking view of the Mississippi as it glided past Gorham. Be it summer or winter, headed east or west, the natural beauty of the area would always be startling. It was easy to imagine the seasoned traveler coming upon this part of the country and exclaiming "Why didn't anyone tell me about this place?!"

Once a thriving city of well over 15,000 in the early part of the 20th century, Murphysboro, as proclaimed by green and white signs posted at city limits, was home to 9900 in 1999. While we approached from the west, we knew a drive into town from any direction would have been pleasant. It was a clean and well-kept town, with straight streets, fine buildings and homes, and a stunning downtown area that had enjoyed a recent resurgence due to what was called the 'Main Street USA' program.

Its library, named after Sallie Logan, was a tribute to the city's history and pride in itself. Sallie was the wife of Thomas Logan, who was the brother of John A. Logan, a Civil War (Union) General and 1884 presidential candidate after whom so many public places (streets, towns, even a junior college in Carterville) in the southwestern portion of the state were named. Murphysboro's library, as well as the courthouse and other long-standing buildings, remained testament to the fact that they could be bent but not broken; they stood firm, tall and proud.

At the end of the 20th century, nine schools served to instruct the young in all regions of the city. Murphysboro High School educated the older kids. Murphysboro Middle School tended to the education of grades six through eight. St. Andrew's School, the oldest school in Murphysboro and only Roman Catholic School in Jackson County, hosted close to 200 students of all ages. Immanuel Lutheran School had well over 100 enrollees from Pre-K to eighth grade, Murphysboro Christian Academy, sponsored by the First Pentecostal Church of Murphysboro, taught almost thirty students. In the lower grades, Pre-K through fifth, public schools were Carruthers Elementary, the town's newest and largest, and McElvain situated in the north end of town.

Two other schools enrolled K-5 as well. They were Lincoln and Logan schools. In 1925, those two held all grades, from first through eighth. And in March of that year, their students experienced a terrifying nightmare that all hoped they would never relive again.

Murphysboro's story, aged somewhat, was a story not only about the children, but by the children, as were most all the stories told to me about the storm. Most of the 1925 Murphysboro tornado survivors were an average age of 83 by the turn of the 21st century. That put them as children in the third grade in the year of the killer storm. It was a rare occurrence to find a survivor able to relate his or her story from the perspective of being older than pre-adolescence. Sometimes it was just as well. The perspective of a child is an undiluted one; 'shock' often cannot be used as an explanation for how big or how horrible something appears to be when relating a tale or experience. The fact of the matter remains that what a child understands as big and horrible most often times simply is.

There was no doubt that what swept across the north and west parts of Murphysboro on March 18, 1925 was both big and horrible. The entire storm, moving forward at 60 miles per hour, made the path from Gorham at 2:26 p.m. and assaulted the northwest side of Murphysboro at 2:34 p.m. In an incredible eight minutes, the storm screamed across railroad tracks, farm sites, rural dwellings, outbuildings, forests and hills—a distance it would take easily twelve minutes or more to traverse by auto at legal speeds even if the road were as straight as the track the tornado carved.

The hills did not faze the killer storm; even Fountain Bluff outside Gorham, which reputedly 'turned' storms, stood helpless. It could do nothing to divert the storm from its direct heading of N 69° E degrees, the path of destruction. The monster storm was almost a mile wide. The wind, with velocities estimated at over 280 miles per hour at that point (as recorded by Stanley Changnon in published reports years later), was leaving a path of splinters and rubble in its wake.

There were no discernible funnels, as had mostly been the case in the mere hour and thirty-three minutes of the storm's life prior to death slamming headlong into Murphysboro (the notable exception being the twin funnels sighted in Biehle). There was only a great, rumbling, black cloud that hung suspended from the afternoon sky.

The cloud moved forward with astonishing, murdering swiftness. It was as if, survivors would later report, the wall of clouds from a normal spring thunderstorm had tilted ninety degrees downward and had chosen to tread heavily across the land at a right angle instead of overhead. For all intents and purposes, it was a vertical horizon, a wall of death closing in on thriving communities with mad abandon.

It was easy to get philosophical about Murphysboro when regarding the extent to which it was devastated by the storm

To Mother Nature, Murphysboro was about to become the proverbial 'red-headed stepchild.' It didn't have bluffs to protect part of it as Gorham did; it was not nestled between hills and hollers as those locations stricken in Missouri had been.

Mother Nature had provided no natural protection to divert death from Murphysboro. To make matters worse, what served as 'protection' for the previously shattered Gorham was very likely what created such trauma for the inhabitants of Murphysboro. Despite the estimates of weathermen and meteorologists, locals had always felt that the possibility of Fountain Bluff having slowed the storm as it slammed across the giant rock formation was a very good one. In any event, it was clear that the storm was on a decelerating course once it left Biehle; and as in any mighty weather phenomena, a slowing of the forward moving speed only served to give the storm impetus to build. Hurricanes are notorious for this characteristic, and this storm was nothing short of a hurricane on land. Wind speeds within the vortex had time to leave the level of fierce and become ferocious; in them, objects could be turned into projectiles that could pierce, dismember, decapitate and shatter flesh and bone with deadly speed and accuracy. Damage could be so complete in a storm of this magnitude that there may be nothing left behind to indicate humanity existed in the stricken area even before the strike was over.

In short, Death would be spending more time in Murphysboro.

To make matters more susceptible to mass destruction, Murphysboro was a young, bustling town. Many new and/or growing businesses were on the upswing, and with such quick growth, corners would sometimes be cut in order to expedite the erecting of structures that might not be fully storm-worthy. Industrial sectors were concentrated in one small, densely populated corner or section of the town that, if stricken, would effectively wipe out the most viable means of continuation of the cash flow that provided revenue for growth in commerce.

It was just such a series of events that caught Murphysboro off guard on that warm mid-March afternoon. Murphysboro was about to experience an apocalyptic bout of bad luck, delivered by Mother Nature herself, and exacerbated by the economic factors that played so serious a role in the level of destruction the town suffered. It left the survivors faced with the task of rebuilding…and of reconciling themselves to the fact that in the big scheme of things, they were victims of circumstance, ill timing, and the basic unpredictability of where the wind was going to blow.

Many of Murphysboro's survivors, the children of the storm, continued to congregate, brought together by virtue of the fact that they had lived through one of the darkest and most extraordinary days of their town's history. They did not all attend the same school at the time of the disaster, but later graduated within the same few years of the late 1930s and early 1940s from Murphysboro High School. Their camaraderie extended not just from alumnus loyalty, but a loyalty to life and to the fact that, unlike the 234 less fortunate in Murphysboro on March 18, 1925, they still possessed it.

> *"We could see the fire and hear the screams. Dad kept saying 'If it gets any closer, we're going to the river.'"—Mary Belle Melvin, age 8 on March 18, 1925*

The Big Muddy River encircled the city of Murphysboro from its origin in the north, which was a convergence of Galum and Beaucoup Creeks, and the Little Muddy River to the east. The Big Muddy continued on around the city to points south until it met the Mississippi.

Mary Belle Melvin, who was 82 in 1999, lived on the other side of the Big Muddy River from Murphysboro. Her home was a comfortable one, built in the years following World War II. It had a breezeway out back through which visitors could enter.

Mary Belle was mobile by means of a wheelchair, and was put there intermittently by a pinched nerve in her leg that had left her weak enough to necessitate some assistance in getting around. And she did that well with the wheelchair. She would often visit with callers in a small library off to the east end of the house.

It was in that library one could find Mary Belle's collection of fans. She had not necessarily been a world traveler, and the fans weren't made in exotic locales (most were crafted in the USA) but to look at them might lead one to a different conclusion. There were delicate, hand-painted scenes on fans framed with what looked like dark mahogany; another such fan was of ivory filigree,

Mary Belle Melvin

cut with amazing precision to give the entire fan a look of lace. Each appeared as though it may have been purchased in an elegant shop in Paris or on more exclusive streets in Hong Kong or Thailand, even though none were.

Books such as *Atlas of the Bible* and *Newnes Complete Needlecraft* lined the shelves of Mary Belle Melvin's library. There were also extensive tomes outlining the world's greatest gems and what to look for in precious and semi-precious stones. Regional history books, as well, caught her eye over the years and she had a section especially for them. My children were fascinated by Mary Belle's collections, and, quieted by the sight (and subsequent examination) of the fans and books, they allowed us to speak uninterrupted.

"I was a librarian at Southern Illinois University-Carbondale for many years," Mary Belle explained as she rolled the chair into her own personal library. Not that being a librarian necessitates nor predicates such an interest in these books; the fact of the matter was that Mary Belle Melvin was a learned woman. Highly intelligent and well-educated, she admitted, and modestly, to having "many degrees." Those included degrees in education as well as being a graduate of Library Science from the University of Illinois in Champaign-Urbana.

But her proudest accomplishment, it seemed, was having kept the graduates of Murphysboro High School's Class of 1939 together and communicating actively.

"My husband and I always took care of the class reunions and of keeping the classes together," Mary Belle said.

Several of the class members became good friends; each of them endured the horror of the tornado of 1925.

It was Joseph K. Melvin, who passed away a few years before Mary Belle's interview, who suggested the name of the club for the ones of the class who continued to meet on a more regular basis.

"He said, 'You need to do this (continue to meet) and call yourselves the 'Survivor's Club','" Mary Belle's husband told her years ago. Since then, the group met for many years at least once a month for lunch.

"We've kept the entire class in contact and have written memoirs of the tornado to keep in our yearbook," she explained, and of this she was very pleased.

Discussing 'the storm,' to Mary Belle, was a natural event that she had learned to handle with great aplomb over the many years since. A few years back, personnel lead by Judith McReary of SIU-C came to Murphysboro in order to videotape a gathering of The Survivors and to listen in on their descriptions of what was surely the most horrific experience of their long lives. The video was shown on local television many times since, especially around the days of mid-March when the winds can become unpredictable.

Mary Belle was obviously proud of the production, but had no intentions of commercializing or capitalizing on the tragedy that befell each member of the group when they were mere children.

"It's good that the story is being told in its entirety by someone," she said of my 1999 project. With a healthy dose of reality, she faced the inevitability of the situation.

"In another couple of years there won't be anyone to tell it," she said firmly. "You'll still hear people talk about 'The Storm,' but it won't have as much impact."

And in the same breath, she admits to her proclivity for expressing the story of The Storm in her own words, simply and modestly: "I've talked a lot."

> *"My grandma said to her brother, 'Oh, Gus, you oughta see my house—the roof's practically blown off.' He said, 'Hell, them on the west end haven't even got houses.'"—Margaret Russell, age 9 on March 18, 1925*

Another former Southern Illinois University-Carbondale employee was also a tornado survivor and was a member of The Survivor's Club.

Margaret Russell spent many years working as a secretary at the institution of higher learning, which had been for years a major employer for the southwestern Illinois area. Her work, she explained, was not difficult or mundane, but during her time there, she had the pleasure of bonding with many people, making her job a delight.

Margaret appeared to bond easily with people. She was a white-haired, bespectacled woman of 83 in 1999, with warm, bright eyes, who moved slowly with a cane but not agonizingly so. She lived with her daughter Susanne and son-in-law Bob Ward and their daughter Toni Hargiss in a comfortable home close to the center of Murphysboro.

Margaret was used to the extended family idea and appreciated its value. Her grandmother, Emma McDowell, raised Margaret from the time she was an infant of sixteen months of age.

"She was a brave lady," Margaret commented about her grandmother. "She was widowed at 35 with four kids, got them all grown and gone and then lost my mom, Jessie Clough, when Mom was just 18."

Margaret revealed that all her life, she was told her mother died of tuberculosis. However, she confided, "I think it wasn't that. She died in 1918, never really recovered from having me. I think it was the Spanish Flu."

Tragedy, therefore, was not a new concept for Grandma Emma McDowell. But what happened on March 18, 1925, was about to lead everyone to new depths of understanding such tragedy.

There was nothing, according to Margaret, that was unusual or notable about the morning of that day in March. Washington School, which was situated on the east end of town (and where the empty shell of the building still stood in 1999) was a school instructing eight grades within its walls. The third graders and others were outside that afternoon at around 2:30 p.m. for recess.

"I don't remember that morning," Margaret admitted after 75 years. "But in the afternoon, when we were out at recess, it was *dark*. There was lightning coming down in streaks. They were wide, broad, and sustained.

"And what I really remember about that storm blowing up that afternoon," Margaret said calmly, "was a lot of thunder. It was off to the west, and we were on the northeast side of the school and couldn't really see anything happening in the sky. But we could hear it." Margaret contends that there was no clap, no rumble. It was just an ongoing, tumultuous series of booming cries of thunder, each more resounding than the last. Her next comment summed it all up in the very words that an eight-year-old third-grader would use:

"It was very loud and very frightening."

> *"There were very few houses left standing on the north side of town. It was just a heap of debris, mainly"—Bernell Howard, age 9 on March 18, 1925*

In a pleasant condo just a couple of blocks off Walnut Street, Bernell Howard, also age 83 in 1999, a widower of recent years, lived alone. His condo had walls of soothing blue; identical color could be found everywhere throughout the house, in the furniture, the wall hangings, the rugs. He seemed to be a content man, and seemed to have had a good life, having worked for the City National Bank in Murphysboro for thirty-seven years, a job, he claimed, he really enjoyed. Prior to this employment, Bernell spent a little over four years in the military (Air Force), "mostly after World War II was over," he said.

"But I did manage to get in before Japan got involved," Bernell said.

His time in the service was spent primarily in Brazil with the Brazilian Air Force, and in the South Atlantic and South Pacific, photomapping.

Bernell's parents had moved to Murphysboro in 1909 from Cape Girardeau, Missouri, and his father had been working at the Murphysboro Shoe Factory, an industry that was developed through the Brown Shoe Company, headquartered out of St. Louis.

Bernell, an only child, attended school with Mary Belle at Lincoln and was also in the third grade on March 18, 1925. His home was directly across the street from Lincoln School, which was located at the intersections of Clay, South 20th, and Dewey Streets on the southwest corner of town.

"There was nothing unusual about the beginning of that day," Bernell said. "We were out at recess as we ordinarily were at that time of day."

However, Bernell indicated, he maintained the conviction that his school officials had *some* kind of advance warning—even if it were only the swiftly moving clouds to the north of that area of town.

"They cut the recess short and got us all back into the school building."

"All the kids...were running down the stairs, and all the nuns were having them pray as they were running"—Laura Marie Miller, age 8 on March 18, 1925

On the west edge of town, in a pretty little subdivision, Laura Marie Miller lived in her house that she said was too big for just her and her dog. Yet there she lived, content in her hometown, surrounded in her big, ranch-style home by mementos such as family portraits, framed exquisitely, and various statues and icons indicating her religion of choice—Catholicism.

Laura Marie, in the early 1920s, attended St. Andrews Catholic School, which had the distinction of being the oldest school in Murphysboro. It was

Laura Marie Miller

the only Catholic school in the whole of Jackson County, and drew in numerous Catholic parents from outside the area intent on having their children experience schooling in the manner of their belief system.

Another item Laura Marie had in her possession was something a little less heartwarming but no less important to her than many of the items that sat framed or in the form of statuary around her house. It was a death list; the beginning of the reporting of the dead of Murphysboro by the local newspaper at the time, the *Daily Independent*. The date of the list, which, after many years, was deeply yellowed and cracked at the edges with age, read March 20, 1925. It was incomplete. Not all of Murphysboro's 234 men, women and children who died in the tragedy were listed on the copy that Laura Marie had sequestered away for so many years so successfully. She carefully spread it out on the table and regarded it as she admitted she didn't remember the day as clearly as she may have in years past, because, after all, "I was just a kid!"

But Laura Marie did recall that the day was warm in the hours preceding the storm. As with the other three of The Survivors interviewed, her third grade class at St. Andrews was out at recess at around 2:30 that afternoon.

"We weren't aware of the storm," she said.

But what she was aware of was the fact that her class was being called in from their recess, their favorite part of the day, early. It made an impact in that it was unexpected and rather unpleasant, to be called inside early from the day's best activity.

"And then there was a big bang," Laura Marie recalled, softly, "and all the windows of the school blew out at the same time."

*   *   *

Back at Lincoln School, the students were experiencing the identical situation.

"What I remember in the classroom was all of a sudden the windows being blown out and the teacher gathering us to the door," Mary Belle Melvin recalled.

But that was not all she remembered of that moment or of the moments that followed. What came next was something perhaps only a child would recall as important, even in light of the situation that the specter of death was looming in the southwest. The innocence of the memory was heart-tugging.

"With all that going on, I remember seeing a picture on the wall of a boy, and he was holding a rabbit in that picture. It was waving back and forth; clacking against the wall of the schoolroom," Mary Belle remembered.

The glass in the frame was on the verge of shattering like the windows of the school had just done.

Mary Belle Melvin commented on the distinct and somewhat haunting memory she harbored of this portrait in the storm.

"The teacher had one of the boys take it down so it wouldn't break in the wind."

*   *   *

"One wall on the second floor fell out," Bernell Howard said of the damage done to his and Mary Belle's school. "Third grade there was on the northwest corner on the first floor," he said. That was allegedly the safest spot in such a building in the event of a storm approaching from the traditional southwest direction.

"They got us on the inside wall," Bernell said.

It was, by far, a preferable place to be in this particular storm. It was sanctuary; it was safe.

There were no injuries in Lincoln School that day.

*   *   *

"They'd called all of us in," Margaret Russell said, "and we had no sooner got in our seats than they said 'Go to the basement!!' There was at least twenty to twenty-five kids per class," she noted, for perspective.

So with eight classes in the school, that added up to a lot of scurrying and panicked, blind running.

"One class was down there (in the basement) having domestic science class," Margaret said, smiling. "Do you know what they call that class today?" She laughed. "Home economics!

"We were down two flights of stairs when I got to this door to the outside. And there, my first thought was to run to my Aunt Mary Heininger, who lived next to the school but a couple of blocks away—we were very close, and she made dresses for me. Then I saw this debris flying fast through the air.

"My teacher, Marian Hammer, grabbed me and took me down into the basement to my cousin, Evelyn Friens, who was in the seventh grade. Miss Hammer handed me to Evelyn. I was hysterical. But Evelyn comforted me. She lived right behind me there in town and was my playmate all those years. Evelyn started praying, and it calmed me down.

"Everybody was praying and crying. There were two stairwells that went down into the basement, one on each side. Pretty much the whole school was by that point in the basement or on the stairs," Margaret continued. "There was a little light in the basement, and I could see how many kids were down there."

*   *   *

On the opposite end of town, at Lincoln School where Bernell Howard and Mary Belle Melvin were, there was evidence of both direct and indirect effects of the giant windstorm that was ripping up the town north of their location.

"There was a grocer's across the street from Lincoln School, and we were looking out the window at it. The grocer had a Model T touring car…and he'd been in it, and he jumped out of it when the winds came!" Bernell said. "Then the wind started blowing the car backwards and a telephone pole fell on it—and stopped it."

*   *   *

At St. Andrews Catholic School, the scene was utter chaos.

"We ran to the hall," Laura Marie said. "There were about 25 kids in each classroom, and it was a two-story building, and all the kids from upstairs were running down the stairs.

"And all the nuns were having them pray as they were running," said Laura Marie Miller.

"I was scared to death."

*   *   *

"It was over so quickly," said Margaret Russell.

*   *   *

"We had no premonition, no prior warning," Mary Belle Melvin remarked. "But since then, you think about it, and put it all together—a muggy day, out at recess, the bells rang early…" Mary Belle shook her head. "We didn't know what it was. We had never seen a tornado before. And no one told us what had happened afterward!

"There were eight grades in Lincoln School. Thirty-five to forty in most classes, and no one killed in Lincoln and Washington schools. You didn't pay any attention, didn't know what it was or that you were upset. We were never given any counseling—it's a wonder we grew up to be normal people. We were upset…but we just…went along."

*   *   *

"My grandma came to get me," recalled Margaret Russell, "I don't remember how this came about, but the teachers instructed us to wait in the school until the parents came for us."

*   *   *

Bernell Howard's story had a similar turn.

"My father came to school to see about me and I met him halfway," he said. "I don't recall whether they kept us corralled outside the building after the storm or let us go. But I remember seeing my father coming to get me.

"He'd said that the storm had taken off the top floor of the three-story shoe factory building."

*   *   *

"We were sent home after it settled," Mary Belle remembered in a somewhat different take on how students were handled at Lincoln School. "We were told to watch for wires; I was trying to walk home the way we usually did; my brother went with me usually, but he wasn't there that time. But I remember this big tree across the sidewalk." That big tree was only the first of many horrors revealed to the schoolchildren.

"I saw a shoe on a leg sticking out from under the tree," Mary Belle said.

Whether it was the shoe that was etched so firmly in her mind or the fact that the leg with the shoe on it was there under the tree, and not another thing could be seen, remained unspoken in Mary Belle's recollection of this first sight of death. She was not fully aware of whether the leg was dismembered, or if the body was still attached, crushed under the weight of the vast uprooted tree. All she was able to render of the recollection was that "I got home by myself, which was very unusual. The roof was off the house, the windows were out, and it had rained pretty hard by the time all this had happened."

Of the many things Mary Belle Melvin remembered of that day, the recollection that she arrived to find the interior of her childhood home drenched from the downpour of rain that accompanied the exit of the massive storm was foremost, something that would naturally have an impact on a child of her age. There was nothing she could do about it, and no one around to help her deal with it in any way...or tell her that all was going to be okay.

*   *   *

"Grandma thought it was terrible, the damage," said Margaret Russell. "She wasn't nervous or upset when she came to pick me up, just grateful, bless her heart; she'd lost so much in her life."

Margaret was very understanding of her grandmother's predicament.

"I'm sure it was frightening for her but she was tough."

However, Margaret revealed the extent to which the storm, which, luckily, spared Grandma injury, affected her.

The demolished landscape that schoolchildren were faced with navigating at the end of the school day as teachers sent them home telling them to watch out for downed wires and lines.

"She said 'Oh, Gus' (her brother) you oughta see my house, the roof's practically blown off.'

"He said 'Hell, Emma, them on the west end haven't even *got* houses.'"

*   *   *

"Some believe that there were two tornadoes in this storm, and they came together around Gartside and 21st Streets," Bernell Howard explained. "There was a vacant piece of ground with enormous big trees and it just broke 'em to pieces—twisted them all together."

Bernell continued.

"We could see the clouds, the storm, as it went out of town, but we were on the edge of the tornado—the major damage was in the northwest part of town. It went across North Ninth Street and just made a path."

*   *   *

"I remember that the priest came over and said it was a bad storm and there are wires down," Laura Marie Miller said.

Damage in an unidentified neighborhood of Murphysboro that was spared the worst of the storm's ferocity.

Laura Marie's instructors were allowing the students to go home if they felt they were going to be able to make the trek. If their homes were close to the school, or they had supervision upon leaving, the students were let go to make their way as best they could.

"I had two older brothers in school, Charles, who was eleven, and James, thirteen. We lived in the north part of town, not too far out. Nowhere was it demolished, though we did have some damage," Laura Marie said.

"But on the way home, we were watching out for wires and things that could hurt us, and we were looking at everything so torn up. Oh, everything was gone. The Christian Church was demolished. I tried to get my brother to hurry. But he kept saying, 'Oh, look-it at that stuff, looky there!'"

Though they were stepping carefully and watching with even more care, "We didn't see any downed wires," Laura Marie said.

"We were just anxious to get home."

*   *   *

"My dad worked out of town as a brick mason, so he wasn't there when it hit," Mary Belle said. "My mom was the one who took us to a little German church straight across from our house. We went there to spend the night."

Mary Belle had in her possession a photo of herself and her brother standing outside the German church across the street.

"The tornado happened just before Easter that year. And when we left our house, which was really in pretty bad shape, my mom said 'take something with you from the house.' She wasn't real specific on what we should take so I took my Easter dress, it was the only thing I could think to take from the house that wasn't wet," Mary Belle said. "Grandpa and Dad came from East St. Louis to town later that day. We all spent the night at the church."

Naturally, there were oddities that remained behind after the winds receded; Mary Belle Melvin took note of these as she recalled the hours following the storm, and the dawn she saw and remembered. She initially skimmed over the darker hours just before the dawn; it was evident that that time period, for now, was too much to talk about.

"Dad had chickens fenced in by the house. That next day we saw the rooster stripped of every feather, crowing…funny what things you remember."

The odd and unusual, and somewhat amusing, were constantly in view immediately following the storm but for Mary Belle, she remembered quite clearly the bizarre.

"There was a big tree out in front of our house that wasn't blown down. For quite some time, in fact several weeks, there was a pair of women's underdrawers up in that tree. They were purple and had double elastic on the legs.

"There was a sycamore tree in our back yard and across from the alley there was a small bungalow. The wind had tossed the roof of the bungalow against the tree, and had laid the four sides straight down, so what did they do later? The people who owned the house pulled the walls up, fixed them together and put the roof back on that bungalow."

Memories like these bordered on the comical as far as Mary Belle Melvin's recollections of the moments and hours following the storm. But there was something else etched deep in her memory that didn't fit in the category of humor—more of human terror than of anything else. And Mary Belle took awhile to divulge the undeniably grim information, but finally, it came.

"You know, when we stayed in the church that night which wasn't too far away from 17th Street, everybody was up keeping an eye out for the fire," Mary Belle revealed.

What was left of the demolished Mobile & Ohio Railroad rail shops in Murphysboro after the wind, then fire, swept through.

To the eight-year-old Mary Belle, the fire, the second wave of disaster to threaten to devour Murphysboro, was a savage creature stalking its prey.

She was painfully aware of it, even though, she said, "I wasn't watching for it; but I could hear it."

The church was in close enough proximity to the burning areas that Mary Belle was able to make this comment:

"I could hear what it was doing."

     *   *   *

Laura Marie Miller remembered vividly the terror that the flames were about to bring with them.

"That night the fires came up from down the hill. My aunt Molly Raddle lived on the west end. She said 'We'd better pack up some stuff and get across to Mt. Carbon; the fire's gonna take the town.'

"But," Laura Marie said, smiling, "she went back to her house to wait out the fire."

*   *   *

"We went out in the yard and saw the fire," said Margaret Russell.

"It looked like the whole town was going to go up. It must have burned all night and into the next day."

Fire is one of those primal prerequisites for the survival of humanity, just like rain and wind. But put all these things together in a form that threatened humanity, as it was in Murphysboro on that day. Perverted, the necessity of fire, and the rain and the wind, made the potential for disaster unimaginable.

The redheaded stepchild that was Murphysboro had been pounded by Mother Nature's hail, lightning and winds in excess of 250 miles per hour; in a matter of minutes, a swath of the northwest side of town was literally obliterated. The insult to the injury was that what was left in shambles was now burning, coal stoves having ignited splintered timbers as if they possessed match heads.

Part of Murphysboro, in the wake of the nightmare twister, was now burning many of its residents *alive.*

"My great uncle Gus Staudt was a machinist at the Mobile and Ohio Railroad Yard downtown. He worked on the steam engines at the engine house," remembered Margaret.

"Now, when he heard the storm coming he'd gotten under his machine and it saved his life, but he got out quick because he could see that along the block of 17th Street a lot of what was left was going up in flames.

"He came back to us at our house on Walnut Street, which is on the east end of town, one of those big houses up on the hill across the street from Tweedy's Convenience Store."

Gus Staudt had gone back to his sister Emma McDowell and her granddaughter Margaret and with great urgency had warned them of the fire that was racing just behind him, stalking him as he ran to their home.

Not knowing how quickly the flames were going to spread, he told Emma to get packed up and be ready to go across the river to Mt. Carbon where they had relatives. His further intent was to go back to the 17th Street area and see what he could do to help once he had assured himself that his sister and grand-niece were going to be okay.

Emma McDowell responded with similar urgency, but in her haste, made some decisions that were questionable; even comical, despite the circumstances.

"I had a little red wagon and my grandma had taken the mattress of the feather bed and rolled it up and put in that wagon," recalled Margaret, smiling. "The bird cage with the bird in it was on top of that. That was what she was going to take to Mt. Carbon. I don't think she could think clearly, it was so frightening.

"But we were able to have laughed so much about that in the past, that bird cage on top of that feather bed in that little red pull wagon."

Margaret Russell and her grandmother waited nervously for hours for uncle Gus to come back and give them the word that they were to head across the river.

"My uncle and some of the men had been fighting the fire all afternoon. About midnight they came back to the house to rest and to get some food. They were covered with soot; they were just dirty. They were worn out. They'd been fighting it all evening.

"But they said they'd got it contained."

*   *   *

"The fire was stopped on 16th and Walnut, but it went through the alley on the 1600 block where Tippey's restaurant was," said Bernell Howard. "There they started dynamiting to blow a path to stop it."

Bernell was haunted by memories of the fire but something more horrific stood out in his mind; an incident in which the urgency the fire brought with it actually created.

"There was a woman trapped in a home on 21st and Hortense Street," he recalled with some difficulty. "They had to amputate her leg to get her out of the building because the fire was coming. Her last name was Raymond. She was around 30 years old."

Bernell shook his head sadly as he said, "I can't remember anything else about her—I just don't know any more."

*   *   *

Mary Belle Melvin was straightforward about what went on the night that the fires swept through the center of Murphysboro. Even though it was difficult for her to talk about it in detail, she spoke very pointedly about what she heard as a result of those flames—and she didn't say much else.

Within the walls of the German church, mere blocks away from the burning, "We could see the fire and hear the screams. My dad said 'If it gets any closer we're going to the river.'

"But it didn't get any closer."

Mary Belle Melvin was a mere eight years of age in the spring of 1925. She said of the trauma that 8-year-old child had experienced, "It got very quiet after a few hours when the people who were fighting the fires got it stopped to a point where they knew it could burn itself out. But there were still several people trapped in the basements in some of those buildings on 17th and 18th streets and that night when it was quiet we could hear them screaming as they burned alive. Nobody could get to them; the buildings had collapsed over them.

"And I wanted to, but I knew I couldn't, do anything to help them."

In the aftermath of the storm, fire in Murphysboro claimed almost as many lives as flying timbers and debris. At the Blue Front Hotel, which was hard hit by the tornado itself, 18 people were burned to death after they took shelter in the basement. As was the case in many homes and businesses that suffered in the storm, a coal-burning stove fell through the floor to the basement when the floor collapsed in the pressure of the winds. Coal burning stoves quickly ignited debris, and those trapped beneath the debris had no way of escape either from the rubble or the agonizing flames.

Mary Belle Melvin was privy to those screams in the evening hours following the worst disaster ever to hit Murphysboro.

*   *   *

Bernell Howard was more stoic about how the fires affected him as a child, though his experiences were certainly no less traumatic. He recalled several instances of the threat of the fire to himself and his family, even though they were in a different direction from where Mary Belle was staying and at a bit more of a distance.

"In an alley a fire truck from Herrin got mired down and burnt up," he said. "A lot of people were trapped, like that Reynolds woman and couldn't get out! There were very few houses left standing on the north end of town. It was just a heap of debris, mainly."

*   *   *

With the M & O R.R. shops destroyed the railroad decided not to rebuild devastating the city financially.

And Laura Marie Miller, whose aunt Molly Raddle warned their family to pack up and get across the river, stayed with her family in their home in Murphysboro that night despite the threat of the flames.

"Our windows were gone and our roof was gone, but we stayed in our house. Dad boarded up the windows and it was cold, but the fire didn't get us and the next day we went to Grandma's house seven or eight miles north of Murphysboro on a farm in a little community called Vergennes. The fire didn't touch us," she said.

*   *   *

The story of Murphysboro was a devastating one, typifying the unholy terror the storm brought with it to the city limits and beyond.

Of all the towns, villages or communities affected by the mighty winds of the killer storm, Murphysboro suffered greatest in many ways. Monetarily, the destruction had by far and away the worst totals in Murphysboro. In 1925

The wind tossed rail cars as easily as it did entire houses during the March 18, 1925 storm.

dollars, property losses, residential, commercial and industrial, amounted to a staggering $10,000,000.00. The Mobile and Ohio (later to become the Gulf, Mobile and Ohio) Railroad machine shops, which had their headquarters in Murphysboro, and their roundhouse were hardest hit, first by raging winds, then fire. According to Margaret Russell, whose uncle Gus Staudt was employed by the railroad, the shops never rebuilt and management made the decision rather quickly to pull out of town. Other businesses, such as Tippey's Restaurant, survived by being relocated to the corner of 16th and Walnut. Although it needed completely rebuilt, owners stayed loyal to the town. Others never recovered.

In the wake of the storm, many more people than in villages stricken before or after Murphysboro were killed by either the force of the winds, the collapse of buildings, the striking of objects flying at speeds faster than bullets, or from fires that raged throughout the industrial areas of the city. The disease that ensued following the exposure to injury or infection took many more lives in the days and weeks that followed. There were a total number of 234 robbed of life by the eight minutes that changed the face and form of Murphysboro forever.

The bone saw and, for the more fortunate, the scalpel's blade, took away much from the living in the form of amputations that weren't already performed by the unique surgical techniques of objects flying in the violent, unheard-of winds. The killer cloud that showed neither mercy nor remorse for its victims left a total of 623 maimed, injured, or harmed in some way. A vast 41 percent of the population of the thriving industrial town was either injured or killed by the tornadic winds that left their indelible mark on the future of Murphysboro for generations.

The nightmare, however, was not over. It was only beginning. With much of west and north Murphysboro left in rubble, with many of the town's residents clinging to life and many more clinging only to hope in the face of having lost everything but the clothing they wore, and with the city's schoolchildren scarred and numbed and confused about what had just happened to them and what lay ahead, the storm unceremoniously moved on.

The twister left behind death and destruction on a scale never before seen; ahead of it lay more, so much more, that the date March 18, 1925, still strikes fear into the hearts of many.

Chapter Seven

DeSoto, Illinois

Mere minutes outside Murphysboro on State Highway 149 lay DeSoto, Illinois. There was nothing exceptional about the drive between the two communities but for the fact that it took about another nine to ten minutes to traverse the distance by auto. The land was pretty, a basic Southern Illinois pattern of fields of crops or cattle, mixed at the edges with peripheral fencerows of trees. A creek bisected the highway a time or two, but essentially, the landscape had only fields and forests to offer before it gave way to another city limit.

In the fading light of the Fourth of July 1999, my children and I could see parties of people, banding together in groups of five or six to two dozen, preparing for their own version of celebration. The city park, which was an obvious point of pride to local DeSotoans, was bustling with softball players, barbecuers and firecracker lighters, scurrying about in an effort to create order in preparation of the big night.

My children and I were not only in search of survivors, but of the dead as well. The date March 18, 1925, was as firmly etched in our minds as carvings in granite, and it was for granite and other solid stones that we were searching. Fascinated by the fact that the death toll of this singular storm was the fifth highest on record as regards overall effects/death tolls for a natural disaster in the United States in the 20th century (behind 1, the Galveston, Texas, Hurricane of 1900, death toll 8,000-plus; 2, the San Francisco Earthquake of April 18, 1906, death toll, 3,400; 3, the Great Hurricane of Okeechobee, Florida, September 16/17, 1928, death toll 1,836; and 4, the Great New England Hurricane of Sept. 21, 1938, death toll estimate 720), we wondered aloud where all the dead had been laid to their final rest.

A tour of each local cemetery was not only time-consuming, but also repeatedly futile. The temperature was so miserably hot, the atmosphere so muggy, that it was difficult to do the 'cemetery crawl,' as we took to calling it, by any method other than from the confines of the sport utility vehicle we drove. So from the windows we'd watch for that date to appear sunstruck across granite or marble or some other finely chiseled material. Upon finding one, (and it was a rare occurrence) we would, all four of us, spring forth from the truck and run to the stone, gazing upon it with wonder and reverence and not a little morbid curiosity—this was inevitable; we were in a cemetery, after all. What had this person been like? Had he or she loved someone? If it were a child (and there were many), how many years of his or her young life would they have in their memories should the resurrection take place today?

DeSoto was distinctly different, we discovered, than any other town or village we had visited so far, as far as cemeteries went. DeSoto, being the small community it was, had one memorial graveyard, a wide, well-kept stretch of land tucked away in the southwest corner of town, and not difficult to find. Our search was fruitful almost immediately. Toward the back, short headstones in rows bore the names of many of the schoolchildren lost to the killer winds and the devastating conditions those winds created.

Three poignant stones in particular bore names of young daughters, whom we assumed, upon first viewing, died when the DeSoto School collapsed upon them. Their names were Barnett—Tina Mae, Marie, and Elsie. Their markers, short, rectangular slabs less than two feet by one foot in dimension, and eight inches high, lay toward the middle of the cemetery. They were almost obscured by the more elaborate, prominent headstones all around them. What made them stand out was the fact that while the date March 18, 1925, repeated within the area in which the bodies of those children were laid to rest, these three daughters had only a birth year and a death year. There was no 'March 18' beside the '1925.' The other outstanding effect was that they were in such close proximity to each other it appeared as though the three might be within a large, common grave. So it was within context that we surmised these daughters had died in the storm—the context of their location within the cemetery and the fact that all three had died in the same year, despite the differences in years of birth. It was a consequence of the engraved image. No matter what the birth date was, the death date remained and cried out in a silent voice of granite over and over and over and over...

Upon seeing these stones, I threw the truck into park with a shout and braved the 7 p.m., 95 degree heat, fifty percent humidity and swarms of mosquitoes and no-seeums to rush to be by the three deceased children's gravesides. My own three very lively children joined me.

The DeSoto School, built in 1910, was a fine building residents were proud of.

They didn't need to be told what this was or to be silent in its presence. The older two could do the math and know they were gazing upon gravestones of children who were their age when the monstrous tornado caught them in its explosive grip and squeezed the life out of them as the village of DeSoto came down around them. My children's silence was emphasized when they both knelt down, along with me and with their three-year-old baby sister, who didn't quite get it, and touched the dark gray stones with flattened palms. With the oppressive heat of the past several days, we expected the stones to be ready to fry eggs. But the sun had already bowed behind the tree line to the west of the cemetery, and no direct sunlight had been upon this granite for at least an hour; no doubt being so close to the ground caused these gravestones to cool sooner than expected.

We ran our hands along the polished section of the stones, across the engraved rectangle in which the years and names sat, and finally over the names and years themselves. Reverence was not an appropriate description of what we were simultaneously experiencing. After all we had been through on that weekend, after all the stories we had heard and the tears we had seen, there was likely not a word or phrase that could have accurately summed up our feelings, especially in children as young as mine were. We knew the horrific power of the storm. We knew that out of a billowing, amorphous cloud, one that gave no one fair warning, there was a giant cone-shaped monster, a killer twister of great size and proportion at the center. We were not only in reverence; we were awestruck with a combination of respect and spine chilling, numbing horror.

The destruction of the DeSoto School, the only school in the immediate area at the time and thus one of the larger schools, took the lives of a total of 38 children, the most killed in a single building collapse during the storm. Of 69 deaths in the village of DeSoto, exactly *fifty-five percent* were children who died in one school building. This, however, was not the total of children taken by the storm, as many more died torn from their mother's arms…and sometimes, we were going to discover, even someone else's arms, someone else who was trying as best as he or she could to hang on to an infant or child in winds which probably exceeded 300 miles per hour.

The entire village of DeSoto experienced monumental losses. The losses were notoriously documented in one of the most graphic depictions through photography to be found in the path of this storm. Several of these photos hung, splendidly mounted, in the bank building downtown, the former Albon State Bank, at the end of the 20th century and beyond, owned by the Old National Bank corporation.

DeSotoans were not and are not to this day daunted by the damage that was inflicted upon their community and populace in 1925. In fact, they, over all the other towns and communities affected by this monster storm, seemed to have taken their adversity and converted it, turning it around to reflect the pride they have in their community. Outside the village hall and fire station building, there stood a monument erected to commemorate the disaster and the victims thereof. This became one of the focal stones in our constant cemetery search; one that, of all the stones we saw, stayed in our minds in its beauty and simplicity, and because of its location outside of a cemetery.

Engraved on this light gray stone was a black etching of a very ominous, violent tornado. There is text inscribed beneath the etching:

"This monument is erected in memory of those who perished in the tornado March 18, 1925 and in gratitude to those survivors whose courage and spirit refused to let the village of DeSoto be lost."

The monument to victims of the Tri-State Tornado located outside the De Soto city hall.

"On our way out I couldn't tell where I was. To tell ya the truth, I couldn't"—Jeanette Ragsdale, age 20 on March 18, 1925

Jeanette Ragsdale was a resident of the Jackson County Rehabilitation Center on the north side of Murphysboro in late 1999. A large, pleasant facility, it housed many of the aged and infirm from Murphysboro, Gorham, Carbondale and DeSoto. A sizable sitting room was constantly occupied. Residents watched movies on the television and read books from the many shelves that lined the walls. They also enjoyed the chirping and singing of several birds housed in the aviary at the center of the high-ceilinged room.

Jeanette appeared pleased with her surroundings and was verbally appreciative of the care she received from those in charge. While many on staff probably didn't know that she, at age 95 in 1999, nearly 75 years later, was a survivor of the tornado disaster of 1925, they treated her with the utmost respect she had earned with her longevity and her positive outlook on the life she led.

The pleasant outlook was reflected in the way she consistently dressed and handled her own appearance. She was regularly outfitted in pretty pants and blouses in bold colors complemented with black, and accessorized with gleaming chains in both silver and gold. While she had the misfortune of missing several lower front teeth, she in no way allowed it to detract from her appearance. Her light gray hair was kept clean and neatly combed; and she even applied a little makeup to her cheeks and lips, as she had been accustomed to do for so many years in the past.

Jeanette was generally confined to a wheelchair, but she did not seem to mind this minor inconvenience. She also had difficulty with her left hand and arm and had to position it properly with her right hand across her chest in order to get comfortable and sit and chat. But again, the term 'handicap' could not adequately describe any affliction Jeanette Ragsdale might have been experiencing. She lived her life and enjoyed it. She was painfully aware, having survived March 18, 1925 unscathed, of the fact that so many in DeSoto on that day could not say the same.

"My sister Eva and I married brothers," she began.

Joe Ragsdale, Eva's husband, worked in the Truax-Traer Coal Company mine north of DeSoto and had two children with Jeanette's sister; they had resided within the village of DeSoto. Joe's brother Frank Ragsdale was a seventh and eighth grade teacher who worked in several schools in the area, including Elkville, DeSoto and Gorham. Frank and his new wife, twenty-year-old Jeanette, lived in the Grimsby area, a little town situated just south of what is now Highway 149, midway between Gorham and Murphysboro, and quite a little distance away from DeSoto. Unlike their siblings, they had no children, and remained childless throughout the course of their married lives.

Describing the earlier part of the day on March 18, 1925, Jeanette said this: "Oh, it was muggy. Kinda misty. Not too cold."

She had come into DeSoto to see her sister Eva and the children, one of whom was Don, only nine months old.

"I went up there to take Don to town. It looked all cloudy and bad, so I'd put him in a buggy to push him up there. Then Sis said 'No, I don't have time to stop and clean him up, Jeanette,'" she remembered. "And he was all messy, he'd been eating lunch a couple of hours before and she'd just let him mess. So I didn't take him.

"You know, I wonder if Eva kinda knew what was about to happen."

> *"Sam…had run all the way from the rail station…he said, 'You go over to Grandma Tippey's, you can make it if you go now!' Oh, Lord, if we'd just listened to him…"—Lillian Bridges, age 16 on March 18, 1925*

Lillian Bullar Bridges owned a fine little white house on the very south end of DeSoto, tucked back away off busy streets and at the end of a short gravel drive.

In the later half of 1999, at age 90, she was a busy woman. She kept a little room, off the side of her carport, cozy in summer and winter, stocking cords of wood for the large wood-burning stove that practically covered one entire end of the carport room, and in the winter kept it heated almost infernally. Food was somewhat stockpiled in the many cabinets and pantries that lined the remaining walls of the kitchenette, having previously been placed there in preparation for any potential Y2K problems, which, fortunately, never materialized. A small sink sat opposite the wood-burning stove; its purpose, it appeared, was only to retrieve running water in order to fill a large, enamel-lined dishpan, which sat on a table next to the stove. Outside the small building there was, of all things, a rusty but perfectly usable hand pump, out of which poured clear, cool, tasty well water all year round.

In those final years of the 20th century, Lillian Bridges was cared for daily by her nephew, Jim Bullar, son of Lillian's younger brother, Carl. Jim called on her around lunchtime every day, bringing her hot meals sometimes, and making arrangements to take his aunt into town should she need anything. This 'anything' included the mere presence of company, as anyone up in age needed occasionally, and besides her nephew, Lillian still visited with several DeSoto residents whom she had known all her life. Otherwise, their visits into town were to stock up on food and necessities, always mindful, even in beautiful weather, of the fact that Illinois winters can be harsh and brutal, and Lillian always wanted to be prepared to ward off the blows such winters may present.

While Lillian, who was called 'Lala' by all who knew her, could very much be considered a 'survivalist,' there was no doubting the fact that she was indeed a survivor, of one of the most catastrophic events her small village and the surrounding communities could ever have imagined. Her response to an inquiry about her experiences on the day of March 18, 1925, was one tinged with reluctance and yet a desire to tell it all, in hopes, perhaps, that by telling it one more time, it might lessen the severity of its impact upon her.

Her immediate remark, however, was indicative of the reality she knew full well, after many decades, that this couldn't possibly be the case.

"That was the awfullest thing I ever went through in my life," she whispered, tears welling up in her blue eyes when first asked about the great storm.

She pulled herself together right away and cautioned, "I can't tell it all, it'd take me three months or six months steady. So I'll just tell you what happened, and you can sort it out from that."

Lala braced herself and, with an unsteady breath, began her tale of that March day of 1925.

"I was 16, just barely that, and I was home with my mom (Gertrude Bullar, 32) that day. See, she was gonna take my little sister Lola, who was seven, into town to try on shoes. Lola needed new shoes, and that was how you did it in those days, you went and tried on lots of shoes and bought a pair that way.

"So I was gonna stay with the baby. That was my little sister, Ruby; she was three months old. Momma was gonna put her in a crib and I could stay with her that day 'til they got back. Oh, I was so proud I was going to take care of the baby! That was the first time I had ever got to stay home with her by myself."

Lala's father Henry Bullar was a miner in the area coal mines and had been gone since early morning. A neighbor, Benny South, was also a miner and worked in the Hurst area mines to the east of DeSoto. Benny's young wife, Edna, was expecting their first child, literally within a matter of hours, as her pains were coming along regularly and had been since overnight. However, Benny was unable to get out of work that day, and had stopped by the Bullar household that morning in order to secure a watchful eye over his beloved wife via Mrs. Bullar.

"He came early that morning," Lala revealed, this being after plans had been made to take Lola for shoes, "and he asked Mom, 'Would you watch Edna today? She's ready to have that baby.'"

Rather than refuse a neighbor in need, Gertrude accepted Benny's request, and rearranged her schedule to meet Edna's needs. She commissioned Lala to take baby Ruby and go on to Edna's with Gertrude and little Lola. When it came time to go shoe shopping, sixteen-year-old Lala would be in charge of both the three-month-old baby and the expectant mother Edna—to Lala's increasing delight. Then mom Gertrude would be back soon to check Edna's progress before Benny got home. It was a neighborly thing to do in a more neighborly day and time; Lala expressed no regret at having agreed to be part of the arrangement her mother consented to on that day.

Later that day, around 2:35, and just before the little family was ready to leave their home to go to their respective errands, Henry Bullar's brother, Lala's uncle Sam Bullar, came running to the door, pounding on it, wild-eyed and out of breath.

"He was on the railroad working. And he'd seen the clouds coming down out of the Murphysboro area. You couldn't fool those men," Lala commented. "They knew about the weather. And Sam knew that what he was seeing was a very bad storm. So he'd run all the way from the rail station and to our house.

"I'll never forget it. He stood on our front porch and was shaking and pointed to the sky and he said to my mom, 'You go over to Grandma Tippey's and take the children; you can make it if you go now!!'"

'Grandma Tippey's' was not the Bullar's Grandma, but was a neighbor lady whose kindness caused her to acquire the moniker of 'Grandma' to everyone she knew. Grandma Tippey had a storm cellar.

Lala shook her head sadly. It was with deep regret that she brought forth the next words of the tale, as if they were grim reminders that she cared not to witness again, but couldn't bring herself to deny.

"Mom said, 'I can't, Sam, I've gotta take care of Edna.'"

Lala then broke into open weeping

"Oh, Lord, if we'd just listened," she cried, "we'd have barely got out of it, but we'd have got out of it. But Momma, she was good to her word. She'd told Benny, and that was that.

"She said to Sam, 'We'll go to Edna's; she's already having severe pains.' And Sam left, and we went on over to Edna's house."

*   *   *

Twenty-year-old Jeanette Ragsdale had gone on ahead uptown without her little nephew, Don Ragsdale. She had decided to stop at Albon's Store, which

Albion's Store in uptown De Soto was where Jeanette Ragsdale was headed on the afternoon of March 18, 1925, with her nephew, but his mom thought he was too messy, so Jeanette went alone.

was owned by one of the town fathers and first village president (1909-1912) George Albon, and was a general store of sorts. Albon's store was located at the end of the first block south off the 'hard road,' the road that later became Highway 149. George Albon's Albon State Bank, as well as a warehouse for his store goods, was set up in the same location.

"Mr. Albon was my uncle," Jeanette said. "He married my mom's sister. So my other sister Mildred worked there at Albon's Store, and that's where I was when it hit."

'It' was the defining moment of catastrophe in DeSoto's history, before or since. It was a black tower of fear that came spiraling and barreling toward the placid town with no intent but to reduce lumber and full-grown trees to matchstick size rubble and to inflict upon the population great anguish and death.

"Mildred and I and the grocery delivery boy, Milford Buckles (who later owned Buckles Drugstore and Café downtown), were in there when it got real bad," Jeanette said.

Because the windows of the building faced to the east, neither she nor the other two in the store saw the storm actually approaching, as it swept in from behind them.

"But it got real bad," said Jeanette. "It was kind of a roaring noise…some thunder, not bad. No lightning. It just got so dark—yes."

The next indication that something was terribly wrong followed the descending blackness.

"Glass started crashing in on the floor," Jeanette said, "and I said to my sister, 'We'd better go under the counter, there's a bad storm!'

"So we did, and we wasn't hurt, even though a big plate glass window off to the left crashed through on the first gust and then all the store caved in on the second."

By the third gust most of the building was collapsed, or swept away.

*　　　*　　　*

At the home of Benny and Edna South, things were turning equally as ominous for the Bullars and for Mrs. South as she endured her labor pains, which were strong enough to double her over.

"Edna had a big, fine, beautiful house, up the street from ours, and there were all these very, very tall maple trees next to the street outside it. Oh, they were so pretty; we just don't have any trees like that now; they were so tall." Lala remembered.

"And Edna had this big picture window in the dining room that looked out onto the street, and we could always see the big trees from right out that window. She had the prettiest, long flowing curtains on it. Edna was a real good seamstress."

Lala went on to explain the next few, intense moments of time, when the entire world for DeSoto was irretrievably changed.

"Mother let me hold our baby because of Edna's pains, and Edna, well, I think she was confused in all the noise and her pains and everything. One of her rooms had already been blown away, but she went over to that door and said, 'Let's go in this room, I believe it's stronger.' But that room wasn't there anymore."

So Gertrude went to the pregnant Edna and made her kneel down where she was. She held Edna's shoulders, Lala said, in order to keep her down amid the possible flying debris that could come through the house in such violent storms. Lala continued to hold on to baby Ruby, standing by the dining table with the infant in her arms. "It was a big table and really strong, I thought," Lala said. "I don't know what I thought it was gonna do for me…I don't know."

At that point Lala's words poured forth unchecked, the trauma of those next moments being relived through agony that years couldn't erase. Her speech was devoid of embellishment or exaggeration; in fact, a manner of understatement was obviously used repeatedly, as if playing down the horror of it all could lessen its impact in some way.

Again, the impossibility of that was all too painfully clear.

"A puff of wind took the big picture window and curtains out, but I still held onto the baby.

"The next puff came and took me and the baby over the table and out the opening (that was left from the dining room window) and over those big maple trees, and those trees were just bending over and breaking off," Lala said.

The sixteen-year-old found herself with baby Ruby in her arms high above the treetops.

"I looked down, and I saw the tops of those trees way below me, and they were whipping back and forth, and here and there a big top branch would crack and go flying off."

Repeatedly, from Missouri and throughout Illinois, witnesses and survivors report that there were three times that the wind (or air pressure) exceeded itself within the boiling mass of the storm and the straight-line winds that accompanied it. The phenomenon was described in many ways; from Peachy Jones in Annapolis describing the storm in three abrupt stages ("It was like this: It was light and it was dark and it was over") to three 'gusts' to 'whooshes' to 'the air got "close" three times.' Thus, asked if there were a third 'puff' of wind, Lala Bridges shook her head sorrowfully.

"I lost the baby while I was over the trees and when I came down, I came down right in the street…I don't remember coming down. I don't remember anything after being up there above the high tops of those trees and looking down and seeing them breaking off. I just found myself in the street in front of Edna's house," she said miserably.

The likelihood of the abrupt altitude and high winds and pressure knocking Lala unconscious is a good one. Mercifully, she did not remember her descent or her impact with the city street in front of Edna's home. All she recalled was coming around…and her mother having landed in the street right beside her.

Lala held out her forearms across her little kitchenette table, displaying the soft, pale underside skin from the wrist up to the elbow.

"Mom was cut in her arms clean through her arteries and veins and it wasn't just pouring blood, it was spurting. It was all over my clothes; there was nothing but blood from my chest to my knees," Lala said.

"Momma couldn't move much. But she got her head up and she saw me and said to me 'Honey, you've gotta cord my arms or I'm gonna bleed to death.'"

Lala was able to sit up in the street and look around. She saw one of Edna's mattresses had landed behind her mother; and behind where she herself sat, she saw some of Edna's good dishcloths. "Edna was such a fine seamstress," Lala noted a second time; it was apparently something she had recognized through all the confusion and turmoil of what had just happened as the storm moved through DeSoto, a semblance of normalcy coursing through abnormality.

"She had got material and had made stout dishcloths from it, and there they were landed next to me. So I reached over like that," she leaned back in her kitchenette chair and extended her arm behind her, "and got one and shook it out and tore it in strips, about three inches wide, and corded Momma's arms."

The terminology 'corded' was a common one back in the early days of the twentieth century, indicating a similar procedure by which a tourniquet is applied to severe wounds in order to slow bleeding, especially of arteries. "Cording," however, differed from a tourniquet in that it did not utilize a stick or other straight object with which to tighten a twist in the wrapped cloth.

"Momma watched real close, the whole thing I was doing, and said, 'Honey, don't tie too tight, it'll stop the circulation,'" Lala remembered through tears. "She watched so close to see that I was doing it right and never panicked and just kept on talking to me. And pretty soon I had them on her arms," she indicated midway between elbow and forearm, "and the blood stopped spurting."

The fact that Gertrude Bullar was able to observe and direct her own first aid spoke volumes for the intestinal fortitude and sheer will the Bullars and others like them in that day and age possessed; but one other factor made the execution of this procedure all the more amazing.

"I was burnt so bad," Lala explained. "There had been a little coal-burning stove in the dining room that had gone down to embers that day because it wasn't quite warm enough to stoke it up, so there was quite a pile of ashes and embers laying in the bottom of it, and the grille that was around the outside of it was of course heated up. And when the first puff came through the house, the stove throwed the fire all over and it burnt me and my seven-year-old sister Lola, a place on her hip that big," Lala said, holding her fingers and thumbs up in a circle at least five inches in diameter, "and it burned her clean to the bone. She died with a scar still in that spot many, many years later.

"I was burnt so bad that the meat had slipped off my fingers and was just hanging. I don't know how I tore that cloth for Momma, because it was so strong, but I did."

Missing from the account so far were Lala's two brothers, Andrew, 14, and Carl, nine. Both had been in the ill-fated DeSoto School house when the storm had roared in. Their own fates had been unknown and perhaps as-yet unconsidered by Lala and her sister Lola—but not so their mother.

"My older brother Andrew had gone through what was left of the schoolhouse and had dug Carl out of the bricks," Lala said. "He'd found him with the bricks having split his head open in a lot of different places."

Lala closed her eyes and slowly revealed the next blow to her family in that day of many.

"Carl's eye had been popped out of its socket by one of the bricks. He was cut from his socket all the way down to his jaw, and the eye was laying out on his cheek almost all the way down to his jaw, too. He was boogered up so bad.

"So Andrew was running and carrying Carl and he carried him way across town. There was one room left in an old house that had mostly been blown away, and he found an old blanket in that room, and went in there and put Carl on it and came to see about us.

"He ran up there and looked down on the ground at Mother and me.

"She knew everything," Lala swore, "or maybe she could just tell by the look on Andrew's face, but I just think Momma knew everything.

"She said, 'How is Carl?'" Lala said in a voice that reflected Gertrude Bullar's peace and calm, a woman whose teenage daughter could tell, by the lilt in the drawn-out name of the younger son as Gertrude Bullar spoke it, that her mother's heart was breaking over the reality of it all.

"Andrew said, 'He's okay.'"

Lala paused after doing what was undoubtedly an almost perfect imitation, almost a voice recording, of what her brother was doing: knowing in his heart of hearts that he was lying to their mother, who was at that moment losing her life's blood in the street.

"He thought Carl had died. And he didn't wanna tell her."

It is nearly impossible to describe, those changes in Lala's voice when she recounted that moment of the tale, for it is eerie that a person can hear something once and, without ever hearing it again, duplicate the tone, tenor, modulation and bare emotion for the listener. But that is exactly what Lala Bridges did during her interview, those voices etched in her memory for an eternity by trauma and loss, to be repeated to those who cared enough to ask.

All three of them, she went on to say, stood or lay in the rubble of the street where just up the way was the broken body of a man. They didn't recognize him; but they could tell that he had been coming home from or going back to work, likely in the mines. He had been carrying a lunch pail and he was lying in the street where giant branches from the huge maple trees in front of the South's residence had killed him where he stood.

A woman who lived near Lala's grandmother had been taking her bath when the storm hit.

"It blew her house away, so she found an old blanket and wrapped it around her, holding it at her throat with one hand and in front of her with the other.

"And she walked down there and stood there and looked at my mother and me for a little bit.

"She didn't have a lot to say. She was dazed."

It would have been fair at that point for the Bullars to think that things could not get much worse.

"Oh, but it got worse all the time," Lala said, her voice a whisper. "It got worse all the time."

*   *   *

At about the time Andrew Bullar was approaching his mother and sister, Jeanette Ragsdale was crawling out from under an incredibly sturdy store counter with her sister and the bag boy, and all three emerged into a scene of utter chaos—but they were unharmed.

In one of the most breathtaking and yet heart-wrenching photos taken during the aftermath of the storm, the photographer stood atop what was left of the first row of buildings on the first block off the hard road, and shot south. A panoramic lens captured the devastated landscape in the late afternoon hours of Wednesday, March 18, 1925. Flat roofs stretched out before the lens, roofs so close to the ground that the photographer, it was said, climbed upon a few adjacent bricks and stepped directly up onto the roofing. To the right, one lone house still stood, its glass burst from the frames; debris littered the lawn. Nearby trees stood twisted, smaller branches denuded from the larger and likely turned into splinters in the grinding winds. All the trees, it appeared, were snapped off at the top at about the same height.

To the far left, a group of men scuffled in the mud, assessing another frame house still standing but appearing highly dangerous, as it had been twisted and likely lifted off its foundation before being allowed to settle back down hard. Lots, cleared of habitation and ornamentation, stood bare, the contents swept away to points north and east.

And straight ahead, down the center of the lens' view, amidst piles of bricks of every shape and size and color, sat the Albon State Bank vault, with a large, white-faced, black-framed clock proclaiming the time of the killer storm's devastating visit to DeSoto when the power was cut brutally from it: 2:50 p.m.

"Many in the bank had got in the vault," Jeanette Ragsdale said. "The grocery and general store I was in was right next to the bank. And those people were saved because of that bank vault. It was the only thing that was really still standing, really still even intact, in all of downtown.

"It was over before I knew what was happening," Jeanette said. Things were in such a turmoil, such an incredible pandemonium visually, that "On our way out, I couldn't tell where I was. To tell ya the truth, I couldn't."

*　　　*　　　*

Three-month-old Ruby Bullar, the baby sister Lala Bullar was so proud to be babysitting on that March day, was found alive but unconscious out in the street where her sister lay dazed and injured, her mother lay bleeding profusely and her brother stood staring at them, stunned beyond movement.

"I don't know who found her," Lala admitted, "but an old woman and her grown son came up out of a cellar across the street to see what had happened, and someone already had the baby picked up and took her over to them.

"They tried to get that grown man, oh, he was about 40 or 41 years old, to hold her and he took her and threw her in a potato bin. And left her there."

Lala's grandmother, who lived on the other side of town, at that point came walking through the incredible debris up to the disaster area the street had become, and approached the injured Bullars.

"Grandma said, 'Where's the baby?'

"I told her, 'I don't know! It threw us in the air and I lost her,'" Lala wept once again at the memory.

"So Grandma said, 'We'll find her, don't you worry.' And she went looking for baby Ruby."

Lala's grandmother found Ruby in the potato bin of the cellar belonging to the reluctant man and his elderly mother.

"Grandma wrapped her up in a big apron; you know, the kind that they used to wear that was big and long and wrapped around the back, and she took her back to her house, and took off the wet clothes Ruby had on, and Grandma wrapped her in a good, clean blanket," Lala said.

The baby, it turned out, was badly burned as well, just as Lala and her little sister Lola had been, by the flying contents of the coal-burning stove at the South residence.

"So Grandma came back and directed Andrew to take her up to the hard road and wait for someone to come along and help get her to a doctor," Lala said.

In the meantime, when Grandma was cleaning up the baby and preparing her for a ride, hopefully into DuQuoin to the hospital, Andrew Bullar had observed what shape his sister and mother were in, and that the baby was in grandma's good hands. In coming back to the devastated street, he had come upon a house that had been blown away from a block east. Beneath that house was a woman pinned, dead. "There had been four men trying to dig her out," Lala said of Andrew's observation he'd made as he passed by on the way to Lala and their mother. "So he ran back to them.

"He said, 'Is that lady alive or dead?'

"They told him, 'Son, she's dead.'

"He told them, 'My mother's alive yet and she needs help—will you see

Roads leading out of De Soto had to be cleared so the injured could be transported to area hospitals like Lala Bridge's mother Gertrude was.

that she gets to the hard road and to the hospital? We can get the dead later.'"

The 'hard road' was the direct path into town, and what later became Highway 149. The Bullars, no doubt, were hoping beyond hope that several interested, not only in rescuing but perhaps in finding out just what the hell had happened in DeSoto, would come to investigate and assist in the effort to get the badly injured to medical care.

So after making such an astute observation for a 14-year-old, Andrew lead the four men to his mother up the street. Behind her remained Edna South's mattress, and with Andrew's help, they loaded his severely injured mother Gertrude onto the mattress and carried her to the road. Waiting there was an intact, slat-sided vegetable truck.

"Those four men got hold of each corner of that mattress and took her quick as they could to that truck, and then they took her on to DuQuoin from there right away," Lala said.

From where she remained in the road, Lala could see the effects of the storm clearly—there was little obstructing her view.

"Three big gusts and it just swept everything away," Lala said. "It cleaned our trees completely out but for a stump; and that had dad's good suit wrapped all around it.

"It didn't leave a pot or a pan or a bedstead on one spot on our ground.

"So Andrew came back and met Grandma, and he took the baby on up the road, and he waited, but not for very long," Lala explained.

While Andrew and Ruby were waiting for an automobile, any auto, to come along and help them to medical assistance, good fortune smiled down upon them after so much heinous misfortune.

"A man came along in a big fine car," Lala said, "and saw those two sitting on the hard road. He came up to them and told them to get in, had Andrew hold the baby on his lap, and they went driving to DuQuoin. Andrew thought they were going to go to the hospital, but they drove up a long driveway to this man's big home.

"Andrew said, 'Hey, this isn't the hospital.'

"The man said, very kindly, 'No, it's not. This is my home. Now get out and take the baby with you.' And they went up to the door.

"His wife had the front door locked. So when she opened the door, and seen them two kids, she said, 'Oh, Harry, why did you bring those little black children home?!'

"He told her, 'These are not little black children, they are little white children, and they're staying here.' The debris," explained Lala regarding the woman's reaction, "had gotten them so filthy, had ground the mud and the dirt so far into their skin, that they looked colored."

Their benefactor was a gentleman by the name of Harold Lancaster, who lived between DuQuoin and DeSoto. He had a son, it turned out, about Andrew's age, and after the two children were cleaned up, Andrew received new clothes and the Lancaster's personal nurse was called for baby Ruby.

"That nurse stayed the day with her, and they hired another one for at night. There was someone with that baby around the clock, taking care of her and making sure her burns were alright," Lala said of the Lancasters' display of Good Samaritanism. But it went farther.

"Andrew had them go back and get Carl out of that house and Mr. Lancaster took him up to the hospital," Lala said, tearing up once again at the memory of her injured little brother. "Mr. Lancaster saw to it that he had the best medical help. They were able to save his eye. They put it back in the socket and he was able to see after that, well enough that he was able to fight in World War II later."

The De Soto School following the tornado. Crumbling bricks and mortar had taken chunks out of Carl Bullar's skull, but he survived the injuries.

Carl's injuries he received at the DeSoto School were quite severe.

"When they went to get him, they knew he was alive because they could see his heartbeat in his veins on the places where his skull had come away from his head. Those places on his brain moved when he breathed," Lala explained.

"And then there was Edna," Lala spoke of her pregnant neighbor whose labor watch meant the severe injuries members of the Bullar family had experienced.

"Two men came along and they made a pack saddle," Lala elaborated, using her arms to show how the men chair-carried the pregnant woman to assistance, folding one over in order to grasp the other, straight, making half a seat. "They let Edna set on their arms and went to the hard road just as fast as they could. They'd had to let her down a time or two with the pressure that the baby was putting on her, but they got her there and to the hospital quick.

"When they took her to the hospital, they took the baby on out.

"Something had hit Edna in her stomach, and had knocked three or four holes into the baby's skull, each one as big as a dime, they said. The baby was born dead.

"And Edna had a 16-year-old sister who couldn't wait for that baby to be born. She'd come up to Edna and Benny's house daily, just waiting and being so excited. Well, Edna's sister was coming home from school early and she got killed in the storm. So Edna and them decided to bury her sister and her own baby together in the same coffin. She was so looking forward to that baby like everything," said Lala…

"…Oh, you couldn't tell how bad it was.

"While I was lying there in that street in that big old pool of blood, a great big hog had come up to me. Oh, it was a nice hog, so fat. But one side was all black hair, and the other side was no hair at all, just slick as a whistle, like what the skin looks like after it's been scalded and the bristles have been scrubbed off when you butcher them." Lala speculated, "I'd swear that to this day, I don't know how, but that hair on that hog had been burned off him in that storm somehow."

Lala believed her mother's blood, in pools all around her, was what drew the hog to her. Now she was faced with her own injuries, to whatever extent those were.

"Finally, my Uncle Lester Bullar came to get me on that street, sitting in that pile of blood.

"He got me under the arms and said, 'Lala, are you the only one here?'

"I told him yes. He said, 'I'm gonna try to get you down to Main Street. Can you stand?'

"I told him, yes, I've stood two or three times now," Lala said, but in her words, it was apparent that the attempts to stand and continue to do so were unsuccessful ones at best.

"'You put your arms around me and I'll support you,' he said.

"We got far as the Methodist Church and he was bragging on me, stepping over all that junk and trash that was all over the road, and how I hadn't let it trip me up.

"And then he said, 'Listen—every telephone wire is down. The electricity could still be on in 'em. You be careful and don't stop.' And you know, my legs locked and I couldn't take one more step for a long, long time. And he rubbed my legs 'til I could move 'em and we started again, and we got to Grandma's from the north part of town to the south.

"You can't imagine how bad it was," Lala declared with great sorrow. "You *can't*."

*   *   *

Gertrude Bullar begged the men who were driving the vegetable truck to stop in Elkville and get her a drink of water. But her blood loss was so severe,

Lala Bullar and her uncle Lester had to make their way past a landscape unrecognizable and covered in mud and other substances, as the injured lay bleeding in the street and in the ruins of their homes. In these photos it appears that one of the aforementioned substances stained the picket fence that once surrounded this residence.

they would not, and instead kept driving into DuQuoin and got her to the hospital.

124

By this point, her son, Carl, their neighbor Edna South, and many, many others of DeSoto were at the hospital as well. And upon reaching her Grandma's house, Lala found a way to get to the hospital herself and arrived in time to see her mother being made comfortable in a private room—courtesy of Mr. Harry Lancaster, who had also seen to it that Lala's father, Henry Bullar, was present with those injured members of his family as well.

"All along, even out on the street, Mom kept saying, 'Honey, I've got to leave you. I'm going to leave you now.' She knew all the time. She just knew," Lala said softly.

"They had her in a private room in a bed facing the wall. I run into that room—I just run real fast and jumped, with my knees on the bed, and hugged her but she couldn't turn but her head.

"Dad got there, and he put his hand up under her head. She couldn't speak when I got there, but when he did this, she looked up at him and said, 'Honey, kiss me, I've gotta go.'"

Lala allowed herself choked emotion, her voice breaking even after almost 75 years.

"He kissed her, and I hugged her, and then she died."

*　　　*　　　*

Uncle Lester Bullar, or "Uncle Lit," as he was called by members of his family, stayed on in DeSoto after the Bullar family was transported to the DuQuoin hospital, making himself available to assist in whatever manner he could to those left behind.

The most traumatic aspect of the destruction of DeSoto remained the schoolhouse, which had collapsed upon children after recess. While many died in the collapse, there were still several outside the building when the storm struck.

"Uncle Lit was worried about Uncle Walt's daughter there at the school, so before he came to get me, he'd run over to the schoolhouse to check on her," Lala said. "He went to the school and saw how bad it was and while he was standing there, just looking at it, he saw something on the light post outside the school. Said it looked strange. So he went up to it and studied it, and then he knew it was a liver, stuck to the light post."

Nellie Bell was almost a teenager, a bit younger than Lala, and was one of those poor unfortunates who was outside and left to face the killer winds unprotected. Uncle Lit could see Nellie's mother, Mrs. Bell, coming across the

A shot of the devastation Uncle Lester Bullar had to pick his way through to check on other relatives at the De Soto school.

school grounds at a distance.

This was when Uncle Lit looked down on either side of the light post and saw the top half of Nellie Bell's body lying on one side, and the bottom half lying on the other side.

He quickly glanced back up to determine whether Mrs. Bell had beheld this sight yet, gathered she had not, and headed off to intercept her before she reached the grisly location.

"The Bells were such wonderful people," said Lala sincerely. "They lived on the south side of town.

"Mrs. Bell was a big woman, really stout, but Uncle Lit put his arms around her, and he said 'Oh, Missus Bell, you don't wanna go over there, come go back to town with me.' And she fought him like a cat, clawed at him and finally got away from him, and then she saw the body."

*   *   *

Many who died in the schoolhouse collapse were little boys. Betty Barnett Moroni, 83, lifelong resident of DeSoto, shed some light on the unusual coincidence, which had turned out to be not a coincidence at all, but a matter of the sentiment in the early part of the twentieth century that girls, being of the fairer sex, should leave the tough work to the boys.

"We were out at recess," Betty, who was in the third grade at the time, said, "but they rang the bell early, because it was getting so dark. We were used to being called in and lining up to go quietly into the building, but that didn't happen this time.

"Our principal, Mr. Stein, told us 'Don't line up! Come inside and go to your seats.'" Betty sighed. "It was so terribly dark.

"As we walked into the classroom, the teacher said, 'Girls, take your seats. Boys, go get the windows.'"

The windows had been open, as they were in many schoolhouses along the path of the tornado from Missouri onward, due to the March warmth. With a storm of any size approaching, the teacher felt it best to keep the windows shut against the elements.

But the elements became the windows as the monster tornado bore down upon their little town.

"Those boys went to get the windows, and it hit. Almost all the boys in our class were killed," Betty said.

The upper floor of the schoolhouse collapsed, the walls literally crumbling, with bricks smashing down on little bodies, crushing the life out of some, knocking holes in others, as in the case of Lala's brother, Carl. Several still outside, such as Nellie Bell, were caught in the wind's violent temper and met fates more terrible than those dealt by the crushing walls.

"See, we had outside toilets," explained Betty Moroni. "My sister Tina, who was twelve, and two other girls were in one of those outside toilets when the winds struck.

"The wind picked up the toilet, and blew it all the way across the railroad tracks with my sister and her two classmates in it," Betty said. "They weren't found until two days later. They were all three of them dead.

"I had another sister, Marie, who was ten, and she was killed inside the school. Another sister, Elsie, was six, and she was killed at home. Our house was blown away; there wasn't a splinter left.

"I thought the world had come to an end. There were buildings on fire, livestock running in the street… I guess the world that I knew had come to an end.

"I remember it as if it happened yesterday," Betty remarked.

Her sentiment was echoed among the voices of the survivors who could still tell their tale years later.

*   *   *

In all, 69 people were killed within the village of DeSoto. As a reiteration, 38 of those 69 were killed in the schoolhouse collapse. The injured numbered 105; the percentage of dead and injured to the village population was a low 19 percent, as DeSoto was a spread-out, fast-growing town in the mid-1920s.

A little girl who was fortunate enough to have not been a victim in the De Soto school March 18, 1925, possibly because she was too young to attend, examines her bike, the front wheel bent, in the ruins of her home.

Monetary losses totaled $400,000 1925 U.S. dollars.

The forward moving speed of the storm was slowing. Engorged with the towns it had swallowed upon crossing the Mississippi—Gorham, Murphysboro, and DeSoto—and choked with debris of every kind including plant, human and animal life, the deadly tornadic storm slowed as it was leaving DeSoto from 60 miles per hour forward moving speed to 56 miles per hour by the time it descended upon its next populated victim. But first it left its mark on a few, new, innocent Illinois communities as it slowed to gather strength, and never, not one time, did it leave the ground.

*　　　*　　　*

As I left the interview with Lillian 'Lala' Bridges, I got into my truck with Jim Bullar to take him back into town, and while I felt strange asking it, I couldn't help but do so and immediately.

"So, her face…" I began, not really knowing where to start but with that introductory phrase. I was compelled; I couldn't help but notice, upon meeting Lala, that there was something very wrong with the features of her face. Her left eyelid was closed over her eye at the outer corner, only slightly more than the right eye, but it was noticeable. On her right cheek, toward her mouth, was a slight bulge. Her lower lip hung away from her teeth on the right side near the cheek bulge in such a manner that when she spoke, the lips didn't close at all. It created little speech impediment, however; she seemed to be used to working with it, indicating to me that she had dealt with it for quite some time. She was, however, forced to keep a napkin in her hand so that she might be able to dab at the saliva as it trickled helplessly from the inside of her mouth to the corner of the open lip.

"She's had a stroke?" I finally asked of Jim.

"No," Jim answered, not at all disturbed nor uncomfortable by the question. And then he dropped it on me.

"Some of it is because of infantile paralysis when she was a little child," he began, "but the rest is nerve damage caused by the fall in the storm," he told me matter-of-factly. "When she said she didn't remember hitting the street, that's probably why; we think she might have landed on her head in some way that it damaged nerves in her face."

I was chilled. Lala was so clear-minded, so well spoken, so ordinary and not at all mentally affected by such a severe blow that might have caused such damage. I couldn't choose my next words delicately or carefully; I was too stunned to care.

"Do you mean she's always been like that?" I asked.

"No," Jim responded again. "It's only gotten this bad in recent years, as she's aged, and her skin has begun to fall a little and be less firm."

I drove and couldn't say any more. The reality of it was screaming around my brain. She had known all these years that she'd had nerve damage to her face, and had lived with it and had made the best of it. And in all of her story, she hadn't one time mentioned any injury she sustained except for the burns on her hands. Her mention of that was only in the context of the fact that she'd had to cord her mother's arms and was having trouble creating the cords. She'd not felt it necessary to call attention to what had happened to her face; she had been far more concerned with telling what had happened to the rest of her family, all of who were so dear to her.

And worst of all, I thought as I stopped uptown and let Jim out at his own truck, Lala was now living daily with the reflection of herself that had to serve only to remind her of that horrible Wednesday of March 18, 1925, and what it had done to her. Over the years the damage was becoming more obvious.

129

What should have healed years ago, in every aspect, was becoming a more obvious wound with each passing day.

I hoped that Lala would enjoy the resurrection that she stated, in the course of our conversation at departure, she believed in so firmly. She deserved the happiness she felt she would receive by being reunited with her mother, and the others who were lost in the storm.

Chapter Eight

Bush, Illinois

Mere minutes' drive outside DeSoto was the city limit of Hurst, where the mining operation that employed so many of those from DeSoto used to be located.

On the other side of Hurst, after a sharp north turn, is the little town of Bush.

The two towns were so close together they had adjoining streets, Hurst on the northeast side, and Bush on the southwest. Why they were not incorporated into one larger area remained a mystery to my children and myself. We found great amusement in the fact that, no sooner did we leave one little burg and turn the corner, than there was another a few yards up the state highway.

A fascinating facet of the area of Bush was a building on the east side of Illinois 149. We first saw it in the twilight of that hot summer night on our way to Ellington. Three evenings later it was no less hot and no less twilight, and the building retained its fascination to us. It was obviously a school building, of dark brick, two stories, all the windows and doors boarded up. There seemed to be no life within or surrounding the building, which rested in a grove of trees and sat on the outer southern edge of Bush. The building, however, spoke volumes with its mute presence. It silently screamed of a time long gone by, when a larger school in a smaller town meant the presence of a better economy for everyone in the community—and this was definitely not the one-room schoolhouse, but the community variety. Its heavy construction and broad dimensions were testimony to that and more.

Questions arose within our minds, the obvious and the inevitable: How many children had it held? When was it built? When was it shut down? Who were the owners now, and what prompted them to keep the obviously defunct structure standing instead of letting it meet its demise by the wrecking ball?

The questions went unanswered. Intrigued by the small town with its railroad crossing and residential homes, but not many businesses to speak of, we drove around in the Fourth of July heat, traveling through the little

Bush water tower

community of Bush.

Our wanderings took us down a road that proclaimed itself to be, by way of the green and white street sign, Cemetery Road. Thinking once again of the dead from the tornado buried in the vicinity, we drove down the dirt and gravel path to the east as a side excursion, dodging deep ruts and potholes and swerves in the road created, no doubt, by four-wheel-drives like ours maneuvering the terrain after the recent heavy showers the area had experienced.

Particularly hideous, almost impassable, was a very low-lying area of road under which there was a culvert, ragged and battered, and a creek crossing through it. The trees here were as thick as the African veldt; the sunset, which provided still adequate daylight, was all but obliterated in the density of the forest surrounding us. It was an ominous, oppressive sensation, traveling this rutted track through the thick shadowing of trees. I could almost envision what the storm of the century appeared to be as it passed over this area at 60 miles per hour. The darkness was complete above our heads; and yet, there ahead of us, we could see glimmers of fading light that, were we to travel quickly and accurately enough, we could reach before it was all gone.

The road eventually leveled out once again, and we came upon an area about which signs present proclaimed to be old mine property. In the glowing golden-red light of the sunset, we came across two young teenagers riding bicycles along the ridge of dirt that had been formed in years past by the mining procedure. And, after explaining what we were in Bush to do, they led us to their grandfather's house, and the stories were underway once again.

Bush was such a small town, such a non-entity in later years, that the fact the tornado struck its northeast edge sometimes escaped researchers. But we were about to find out that, as in all the villages, even the ones that weren't the size of Murphysboro, the twister had had a profound effect upon the residents. They were still in the area; some in town, some having moved away. But the memories, unlike the storm itself, never left them.

> *"The word 'wind' in Italian is 'vento.' There is no word for tornado. They don't have tornadoes in Italy—no storms like they have here"* —Joe Martel, *age 10 on March 18, 1925*

There was a family by the name of Martel in a village in Italy called Cimina. Located just below Rome, the village was a farming community, specializing in the kinds of fruit, grains and vegetables that were grown for sustenance and trade within local provinces of Italy in the early part of the twentieth century.

The Martels were farmers from the maternal branch of the family, an arrangement that was commonplace at the time. Farms often came with the bride as a marriage dowry, and this Martel farm in Italy was one of them. One of the sons of this fine woman proved himself too lazy to continue to work on the family farm. So in 1920, he took his wife and his young son, Joseph, and moved to America, where he settled in a little Southern Illinois community called Hillsboro.

From there, little Joseph went to school, not able to speak a word of English.

"I felt like sinking into the floor," Joe Martel said of his first few days in the country school in Hillsboro, which was situated near Litchfield on the east-central part of the state. "So a teacher came to me and said something in English, and brought out the First Grade reading book with the English alphabet in it. I at least knew that it was an alphabet; I looked at it and read 'em off in Italian.

"She smiled. She had taken Latin in college, and she knew quite a bit of what I was saying, and she took care of me and showed me how to read and to speak in English."

When Joe was ten years old and in the fourth grade, his family, by this time having a little daughter, Joe's baby sister, in tow, left Hillsboro for Bush, Illinois. It was here that Joe Martel made significant progress in school, having come from a country school in which all grades, from first to eighth, shared a room.

"That one in Bush was a real school," Joe said. "There in Hillsboro, there were eight grades in the classroom, and everybody had to be quiet when one row of students had their lessons, and the teacher would move on across the rows, and if you were paying attention, you could pick up on the big kids' lessons. But in Bush, there were classrooms for every grade. All the kids in Bush and Hurst went to school there. It was that big brick building on the west side of the road. Most of the kids that went there were miner's kids; some were kids of farmers, like me."

The mining community was based on the Hurst Coal Mines operation, and was a major employer for many towns and villages in the Franklin and Jackson county area.

"There were 'camps' in Bush and Hurst; there was the 'North Bush' mining camp, where there was housing, and there was either the middle or south of Bush that had a camp called 'New Jerusalem,'" Joe said. "They had a big company store and a camp, toward the school, that the mines had put up to house the miners. They would rent only to the people in the mines. When you worked at the mines, sometimes you didn't even get a paycheck. The miners would borrow on their wages and sometimes they'd overspend; then they'd have to get their rent taken out of their pay because the mines owned the homes that the miners lived in!" Joe Martel laughed. "I worked in the mines in Zeigler long enough to learn a lesson and come up top."

Joe eventually went to school for sheet metal fabrication and had his own business making equipment for big mines around Galatia, Illinois. One of the things he made was mine tipples.

"Some equipment weighs 25-30 tons," he noted. "Tipples weigh thousands of tons."

Joe Martel explained the importance of mine tipples, giant contraptions that marked the presence of a mine below where the tipple sat.

"Coal goes on a belt from out of the ground, up in a tipple, and it's crushed. Then on the tipple belt, it's washed, and the rock is separated from the coal, because coal floats and the rock goes down. Funnels spin the water, and that's how they separate.

"It goes into a conveyor belt to a railroad car, and that fills up. Then the train moves the car, and it goes over a scale (for weight confirmation), and after that it's shipped.

"I have worked for all major mines in southern Illinois, Zeigler Coal for 58 years. For the last five years of it I worked two shifts," Joe said.

He was, he felt, an authority on mines and most of the equipment they used. It was a lesson learned over many years as a responsible adult, tending to a family and carving a career for himself. It was not too far a cry from the life he saw going on around him with the miners in the town he lived as a child.

It was in that town, in the school that educated the children of Bush and Hurst, where Joe Martel began to show that he was an excellent student.

"When I was in that Bush school, I could get any kid in the eighth grade class. I had got to learning the grades in the little country school, and I could take on any lesson that any eighth grade class kid could try. And they weren't as well behaved as those kids in the Hillsboro school. Those kids were well behaved. I never saw a kid get a whippin' in the country school. That Bush schoolhouse was different."

Joe Martel was seated in the schoolhouse on the afternoon of March 18, 1925.

"It was an okay day," Joe remembered with a rather unremarkable tone. "It was later in the day, but we were still in class, and I do remember my row was along the windows on the east side of the school."

Gazing out the east, it was difficult for the children to tell exactly what was happening to the afternoon sky, as the change approached from the west.

"We didn't know what it was. It just got really dark and the wind was blowing, and then it was over."

The schoolchildren were at the verge of dismissal anyway, being that it was just before the three o'clock hour. The teachers and school principal determined to send the children home, and as was the usual case, sent many of them walking, even with the threat of the storm as it moved off to the north of town. Several followed the powerful force they had all experienced but knew not on what scale, when it passed them by in as they sat in the sturdy schoolhouse.

"After the darkness was past, they let us out, and I went walking home," Joe Martel said.

The Union Pacific rail line still ran through Bush, on a southwest to northeast track and back, as it did in 1925.

"Past the railroad tracks going north, back then, on the right, there used to be two rows of houses called 'New Bush.'"

This, explained Joe, was previously a part of the mining housing, or 'camps,' that the Hurst Mine Company had set up for employees. That particular section was now open for other renters. That was where the Martels had established their home.

"When class let out, the wind was blowing so hard still you couldn't stand up. Wind kept blowing steady for a couple of hours.

"When I got home, both rows of houses were all torn up. We lived in the second house on the south side.

"There was five people living next door (to us), a couple of older people, and a mom and dad and I think one of their kids. One of 'em had tried to call Mom and my little sister to come over and stay there with them when they saw the storm comin'," Joe said. "But Mom wouldn't go on account of the rain."

His mother's reluctance to get drenched in the approaching storm undoubtedly saved her life.

"All five of 'em were killed. It blew their house away," Joe said of his neighbors.

Of those killed in the area considered to be Bush, there were seven in number. The large majority, Joe noted, were the five right next door to him.

"There wasn't anything of that house left standing but part of a wall. In fact, I would say that of 14 to 16 houses in the rows, only two or three had any walls standing at all."

Joe said simply, "I don't remember their names," of the people who resided there.

His family didn't fare badly, but they were not spared the wrath of the mighty storm.

"My sister got blowed up in the air, mother too, and it let 'em down near the house. Sis found the house and crawled under the floor, and she stayed there for awhile," Joe said.

His sister at the age of five, no doubt, was in shock.

"But my mom, it blowed her into Bush proper, and it took all her clothes off her but her panties, and a two-by-four hit Mom so hard in the leg that her leg broke. And she about lost the leg later, it was so bad."

The enormity of the storm was not lost on little Joe Martel, who was nearing shock himself at the vague idea of a major storm, one that could not necessarily be seen or felt from the large brick schoolhouse south of the disaster area. It had almost insidiously swept through and obliterated everything he had known, masked by the sturdy walls of the fine Bush schoolhouse.

"If you saw what really it did do, you wouldn't believe it," he insisted. "Railroad axles, heaviest thing there is, was scattered around the rail yard. Two-by-fours were driven through the water towers in Bush. There was galvanized tin wrapped around trees.

"I was kinda young and just was lost when I went over there and saw the floor of our house and nothing else," Joe said.

He did not realize that his little sister was lying in shock under the very floor at which he was staring, slack-jawed.

"Someone told me where Mom was, don't know how they knew, and I went looking for her and found her."

> *"People were just goggle-eyed...I saw bodies laying out in a garage when (we) went walking by there"—John DeHart, age 6 on March 18, 1925*

John DeHart, who was the grandfather of the two bicycling teenagers, Eric and Sarah DeHart, whom we saw at the defunct mine sites, offered his take on the aftermath of the storm—the only part he clearly remembered, being a child of six at the time.

"I lived with my parents down a dirt road outside town, and we were coming into town in our model T Ford," John explained.

"We didn't know 'til we got up to Highway 13 (with intersection at 149) that something terrible had happened. See, Grandpa owned property here in town, and we were coming to visit him...but when we got up here to where the tornado had been, all we saw was junk."

John shook his head and his eyebrows shot straight up, thinking about what the next sight he beheld was. There was a hesitation, and he continued.

What was left of the village of Bush, approximately in the vicinity of the mine housing.

"People were just goggle-eyed. I know my parents didn't mean for me to, but I saw bodies laying out in a garage when they went walking by there, taking in the damage themselves.

"We couldn't miss seeing them. We walked straight through town."

While only seven souls lost their lives within the town of Bush, many of the injured as well were ones who were laid out in the garage that John spoke of, and there were 37 of those.

The DeHart family, however, did not stop in Bush for long, but headed back home, making a side trip of DeSoto and Murphysboro when they heard that the damage had passed through those towns as well.

"People were going crazy in Murphysboro," John said.

It was all he had to say about that town, all he really had to say about the storm and the aftermath he personally observed at the tender age of six. Many years later, it seemed to be too much for him to try to recall the details of such a traumatic sight.

The subject was changed at that moment, and his grandchildren came in, to remind him that they were all in the here and now, and safe.

* * *

Joe Martel was also traumatized, understandably so, at the sights and sounds of the north edge of Bush having come under siege by the elements of nature no one could control.

His memory was better in the aftermath, as it was about what occurred during the resurrection that eventually began within every town in the path in the days following the mighty storm; Bush was no exception.

"We all went to some friends' houses to stay at first, and then, there were some empty houses in town, so we cleaned one up and kinda camped out," he said.

As odd as it sounded, and as illegal, too, this was something that was not uncommon in the days and weeks following the storm. Empty houses, presumed to be either abandoned or left behind in the wake of the death (or disappearance if death was not confirmed) of the owner, were beset upon by the living who had lost everything *but* their lives. They turned the empty shells into much-needed shelter, a habitat, and eventually, a home. To emphasize how important it was to the Martel family to have found this abandoned home, Joe Martel explained, "All we had were the clothes we had on when the storm struck." And this, in the case of his poor mother, was not very much clothing at all.

The Red Cross, the organization whose efforts following the storm was in equal parts praised and maligned, attempted to aid the survivors. They came along not too long after the Martels set up house.

"When the Red Cross came to donate, they throwed shoes and socks and two pair of overalls, didn't ask what size we needed, just like you'd throw something to eat to a dog. I had a pair of overalls I coulda put two legs in, they were so big. And we had to throw the shoes away, *they* were so big." Joe Martel's mother was more resourceful with the overalls, however: "Mom took the overalls apart and made more clothes out of 'em."

The time of the year in which the storm hit, spring, was probably just as good a time for it to strike as any, as there were no crops in the fields yet and the early gardens that were sprouting forth were not heavily damaged in the wild forces of the wind.

"That was about the time to plant seed," Joe said, "so we rented some ground from the railroad, whose owners were also the ones who owned the mine. We raised enough food, potatoes, turnips, and canned everything, so we didn't depend on the Red Cross for anything, like so many people did in the hard months after the storm."

Their resourcefulness, stemming from the family of a man who ended up in Illinois because he was considered too lazy by the standards of his Italian farming family, was nothing short of remarkable—although some might have believed that it was mostly luck.

It was through luck that Joe's mother and sister had not been inclined to go to the house next door when they were implored to do so, luck that his mother lost most all her clothes instead of life and limb, and luck that the vacant house was made available and taken by the Martels instead of someone else first.

"After the tornado we sold a lot of stuff we grew to get money," Joe said.

It was a time of resourcefulness as opposed to desperation; and the Martels, as well as many other families in Bush and in Franklin County, drew upon their reserves in order to ride out the secondary storm the first one left in its wake.

Bush lost seven residents to the storm; other than the Martels' neighbors, neither Joe Martel nor John DeHart knew who the other two were. Thirty-seven people were injured, some severely; but it was a small slice of the entire population, three percent, that was dead or injured. The property losses in 1925 dollars amounted to $212,000; most was damage to the rail property or equipment, or mine-owned property, including rental houses.

The toll on the minds of those who survived could not be recorded by numbers or in dollar amounts.

Chapter Nine

Zeigler/Plumfield, Illinois

Leaving Bush on Highway 149 meant traveling north, on a straight stretch of blacktop if there ever was one, until the turn to the east that lead directly into a small town called Royalton.

Royalton had the good fortune of the twister staying on its N 69 E heading. The town lay just a few miles to the north of the track. Few on that fateful Wednesday knew what the storm that darkened the skies with the debris of sheer horror to the south really meant. There was a constant air of surrender to the forces of nature from small-town folk back then. No one could do anything about the weather, that was for sure, but life went on around it, and in Royalton, in later years with its bars as many in number as its churches, life did just that on March 18, 1925.

The land was literally flat as a pancake in the area between Royalton and Zeigler, and beneath it lay many coal mines. Though in 1999 not much existed to indicate this former industry and how active it was in this particular section of Illinois, the terrain gave indication of it by aged 'slag' heaps of soil and refuse brought to the surface during the process of mining and often left in smoldering piles upon the flat fields. While 'slag' was generally a term given to the production of steel and metal smelting, the fact was that impurities were brought up from coal mining as well, and several piles burned for lengthy periods of time during active mining. The area was dotted with mounds created by the mine working, some due to former slag heaps, some, strip mining, which was the surface form of mining coal, only possible when the seam of coal was closer to the surface of the ground.

There were some large mounds that dotted the surface off the roadway between Royalton and Zeigler, indicative of the surface strip mining, accomplished in more recent years. However, many of the mines in the area were 'deep mines.' These were accessed by deep shafts cut into the coal formations, and further reached by hollowing out underground rooms for the miners to work in, after creating pillars of coal for support.

Coal mining in Illinois, and indeed, in the entire Midwestern section of the

U.S., was on a high roll of $34 per ton in the energy crunch of the late 70s and early 80s. However, Midwestern coal was dealt a death blow with the passing of the Environmental Protection Agency's Clean Air Act, federal legislation in 1990, which required coal-burning electric utilities to lower smokestack emissions, especially of sulfur dioxide. And sulfur dioxide was something Midwestern coal contained a lot of. Coal mines in Wyoming's Powder River Basin, by contrast, retrieved coal that was low in sulfur dioxide emissions when it burned—and it was shipped all over the country as a result instead of Illinois coal.

The industry of mining coal had developed, like most everything else, into something quite high-tech and modernized mechanically. In later years, heavy equipment and machinery took the place of the miner's traditional tools of the pick and shovel in places across the nation where it was still being mined—though in Franklin County, by the turn of the twenty-first century, there were no active mines at all.

It was a sketch of such a miner, equipped with pick-ax, shovel and miner's lighted helmet, which greeted us upon our arrival into Zeigler from the southwest on Highway 149. The miner typified literally thousands of men from the earliest days of the industry to the final raising of the last man on the shaft elevator in later years. His face was hardened, seamed like the coal he sought with his pick-ax and shovel; his eyes were piercing and yet bright, like good bituminous coal lumps sparkled when the light hits them just right. But his skin was pale from hours in the dark and damp of the mines, and his jet black hair, peeking out from the miner's helmet, stood in dark contrast to the skin of his face, neck and arms. His pick-ax was slung over one shoulder, and he used the shovel to lean against on the other side, for he was tired from his ten or twelve hours of mining the black gold of the Midwest. In fact, the hardened set of his face bore the expression of an as-yet-unheaved sigh; it was as if he had been waiting for the artist that drew him to stop the work long enough for the miner to take a great breath, let it out, and continue his pose until the project were done. His clothing was probably as new as a miner could hope for, but dirty all the same. The coal dust and grit clung tenaciously to the fibers of his suspendered jeans and rolled-up sleeves. But he was proud. He was the representative of the livelihood of the towns and communities which we were passing through. And he knew the import of this position.

The drawing of the long, lanky, mine-hardened man stood next to a billboard of sorts which read "The people of Zeigler welcome you to our town" and was followed by a list of community and civic organizations. There were many. Like DeSoto before it, Zeigler was proud of its people and of its contribution to the economic development of Southern Illinois. From its coal mining to its civic groups, it had always been known in the area as a progressive, albeit small, town.

Particularly of interest in Zeigler was the town circle, which was merely a rounded version of the traditional town square. The center of the circle was now a small park, with a canopy of trees and a few benches for relaxing.

In days gone by, when the area mines were flourishing, the 'Company Store' was well-situated at the middle of what was now this modest park. Miners came to the store, locals said, to get their 'duds,' and a credit was given them should they not have the funds immediately available. Then, come payday, the Zeigler Coal Company would deduct any expenditures off the worker's paycheck, and the balance he would take home. Thus was developed the theme around the Tennessee Ernie Ford song, "Sixteen Tons," in which Mr. Ford sang in his deep basso profundo, "I owe my soul to the company store…"

Around the periphery of the circle were several active as well as defunct businesses. In one of the active businesses, known simply as the Zeigler Coffee Shop, I ran into a group of folks who knew the names of several survivors of the great storm on 1925, and were very helpful. However, one in particular, Darlene Davidson, pointed out that the storm, contrary to reports I had received in my initial research, did not pass through Zeigler at all, but did strike and do much damage to a little settlement outside Zeigler called Plumfield. Having never been to the Plumfield area, I took the name of a survivor and called, hoping that, wherever he was when the tornado made contact with this little mining community in Franklin County, he might be able to shed some light on what darkness shadowed that long ago Wednesday in March, 1925.

"How well I remember that!"—Cecil Spain,
age 21 on March 18, 1925

Thus began the conversation with Mr. Cecil Spain, 95 years of age in late 1999, former National Association of Letter Carriers member and resident of

Zeigler, Illinois, for a total of 76 years.

"My version of this storm will be different than anything you've ever heard, not just because all those I knew in it are dead," Cecil insisted, "but because I was outside the path of the storm."

The fact that Cecil was concerned about those he knew being gone was testament to a time gone by. People didn't talk openly about private business. Cecil continued, explaining that there were so many involved in the tale, he was concerned about getting it right according to what *they* remembered as well. It was also as if, to Cecil, the absence of anyone else to verify his story might make it appear less valid, less important, and he was prefacing the tale with a cautionary note. However, the fact remains that even two people, standing side by side and viewing the same horrific winds, will each relate the incident in a different way when asked. When this fact was pointed out to Cecil, he concurred, but he added a further introductory statement: "I'm the only one left who can tell this version of the storm, being on the outside instead of a 'survivor.' Everyone else involved in it is gone."

Cecil Spain put forth the distinct impression of being a highly educated man, in that he was well spoken, with a clear dialect and not prone to slang terminology or lazy speech. However, he insisted that he did not possess a college degree. He revealed that he took advantage of a mail order course in English from a Chicago University back in the early twenties, at the behest of his roommate at the time, who was a physician, a general practitioner in the Zeigler area. Cecil's intellect was one of those typed as innate. He was born an intelligent child, and he had used his intellect to become wise over the years.

His eyes were bright for his 95 years, although his hearing was a bit dim. Even with a hearing aid, visitors had to shout at times to be certain Cecil was able to understand them. He got around just fine for a man about to approach a century of living. With his straight posture, his easy gait and his ability to speak so eloquently about the subject at hand, he could have easily been mistaken for a man twenty years younger, and at least six years more college-educated.

Cecil and his wife of 74 years, Clara, lived in a beautiful little house on the west side of Zeigler. The front of their home had been converted from a beauty parlor that Clara had owned and operated for many years. The parlor fixtures were gone, but there were the traditional beauty parlor mirrors still in place—and all over the house, for Clara Spain loved mirrors, and had them on almost every wall in the home. At 93 years of age, Clara was so beautiful that she had no fear of mirrors, and took great pride in them as she did in her extensive costume jewelry collection, and, above all, in her beloved husband.

"I have lived here since 1923," Cecil began. "I was living in Marion before that, and Zeigler was a frontier town just being built.

"My brother-in-law was living here, and he said 'You'd better leave Marion and come here and make seventy-five cents an hour!'"

Cecil laughed. "I know, I know; it was in the fall of 1923, and that was a lot of money to be made back then, but so much for that.

"I came to Zeigler. My first job at that time was hand-mixing concrete with shovels. I was a strong kid! From there they asked if I could drive a truck. Model T Fords were prevalent. So that was what I did next.

"Anyway, we were going to pave the streets, so I got a job driving a truck; and this paving job started in the spring of 1924. In March of 1925 we were going to pave the road from Royalton to Hurst."

Hardpack dirt roads were common in the early 1920s, due for the most part to the fact that so many still got around via horse and buggy. With the advent of the affordable Model T, however, more and more common folks were switching from the switch to the steering wheel, and the hardpack was almost an inconvenience. Rain and other adverse weather conditions encouraged and prompted paving procedures, especially heavily traveled roads between the mining towns. The area was bustling and growing. The roads needed to keep up with the times. Cecil Spain's employers were there to see to it that it was going to be done.

"We had a big garage on Cockspur Street here in town. On March 18, in the afternoon, we were working on the trucks and getting set to go to Royalton," he said. "There were five or six of us with a supervisor, and we were all doing things like putting in spark plugs, when all of a sudden it began to get real dark."

Cecil Spain sat back in his wooden kitchen chair, tipped it up against the wall a bit, and let it back down easily as he said "Real dark; thunder; ooh, lightning!

"Someone said 'Turn on the lights!' We had the kind of lights that were bulbs at the end of long cords, so we pulled the cords down and got the lights on."

His hands turned in the air, as if reaching for the ghosts of those old light cords and their switches. Cecil looked heavenward momentarily as he continued.

"I saw no clouds."

None of the men in the garage, he noted, were interested at that point in going to the outside of the building nor to the windows to see just what was the matter. They naturally figured a very bad storm was approaching.

"It just got *dark*," Cecil repeated.

"People never thought of tornadoes in those days," Cecil said, and, although it sounded a bit unbelievable, his next sentiment had been expressed over and over by the survivors of this particular storm.

"We'd never heard of a tornado in that time.

"It started hailing, golf ball size hail, round, not too thick," he said. "We just went ahead with our work. We didn't think too much of it."

Cecil paused briefly, as if to let the road crew's obvious disregard of the weather conditions pass as unnoticed as the storm that had only mildly disrupted their workday.

Then he went on to say, "In fifteen to twenty minutes, one of our friends came in from West Frankfort. He worked for a railroad company as a dispatcher.

"'Oh, we had an awful windstorm,' he told us. 'Up at Plumfield, our church is gone! I had to drive through two buildings that had been blown across the highway!'"

Cecil continued, "He said it was a terrible windstorm. But on this side of the Big Muddy River, there was no sign of damage. And what he said about driving *through* a house…"

Cecil shook his head, paused, and then went on.

"It never bothered me at that moment. So we finished our work at four p.m. and I went home."

Cecil Spain shared an apartment in town with two other men from the area.

"At that time there were three of us, Dr. Sigmund Tashma, a surgeon and a general practitioner; and Dr. D.J. McCullough, a dentist; and me. We had an upstairs apartment over the bank building.

"Dr. Tashma had gone to St. Louis with one of our town merchants, and he wasn't in that evening.

"So I went upstairs and the phone rang. I answered the phone, and it was a message coming in from either Christopher or DuQuoin, I don't remember, but it was one of those hospitals, for Dr. Tashma. They were needing as many doctors as they could get in West Frankfort and wanted Dr. Tashma there.

"So I said, 'I'll tell him when he gets in.'"

Cecil continued to relate this part of his story with unwavering clarity as he went on: "Dr. Tashma came in at dusk; I told him and he said, 'Alright, let me get my bag and instruments.' And he had an extra bag, and he packed that one, too."

This was unusual, Cecil said; normally, the doctor didn't take an extra bag on calls, and he didn't seem to be displaying any sense of urgency for this call, but Cecil pointed out that something, unspoken by the doctor, caused him to go ahead and pack an extra bag anyway.

"I'll never forget that, packing an extra bag," said Cecil; it was that unusual.

"He said, 'You drive me over there, I may have to stay there,'" Cecil continued as the doctor's premonition of the matter was turned up a notch. "And we took off for West Frankfort.

"I drove through that house," Cecil said reflectively, "the one the railroad dispatcher told us about. It was a four or five room house in pieces, laying right in the middle of the road." The car Cecil and Dr. Tashma were traveling in could not detour around the windswept house on either side of the road; they simply had to drive *through* it.

"I'll never forget it," Cecil said. "The roof was gone; we drove right over the floor. We *had* to; we couldn't drive around it."

Cecil left his roommate Dr. Tashma in West Frankfort and went back to Zeigler. Cecil did not comment on the conditions he'd seen in the city because there was nothing of note in the part of town where he'd dropped off the doctor.

"In the meantime, early the next morning when I went back in to work with the truck crew, our supervisor Ted met with us and said, 'Everybody get in your trucks, we're donating our trucks and our time and going to haul away the debris in Murphysboro.'"

Given the fact that the darkness that had descended upon Zeigler, followed by hail and reports of a 'terrible windstorm,' had not impacted the garage crew seriously, they no doubt wondered what all the fuss was about.

They were about to discover it first hand.

"We drove out of Royalton, Hurst, Bush, and into DeSoto," Cecil remembered, his voice flat, as if allowing no emotion to escape and color the recollection.

He said little of the damage he observed in the town decimated by the tornado that tore through the heart of the business district.

"I remember a brick house on the highway in DeSoto as the only building still standing. That house is still there, a big, two story brick house," Cecil declared.

The likelihood of this being the house belonging to George Albon, city father of DeSoto, was a good one. So many, it was reported, were able to board in Albon's big brick structure that survived the storm, noting it was one of the few that did.

"Being a young guy, you didn't pay too much attention," Cecil said of himself and the work crew upon their arrival in Murphysboro. How they felt about the situation, and the level of destruction the city had experienced, was something that remained mostly unsaid by Cecil—and that said volumes about the incident as a whole. They simply, Cecil remembered, went about their business. They were there to do a job, and that was what they were going to do.

"We go over there, and those fellows there would load onto our trucks. Then we'd drive out and dump the wood and bricks in a pile. We had dump trucks; we'd just back them up and dump out the load, and go back for more."

Regarding the fire that consumed so much of Murphysboro, Cecil noted only that "We saw smoke coming from the fire downtown—just smoldering. There might've been some blaze here and there, but we were more worried about getting tires punctured in all the debris than about burning up.

"We spent the day there. Had our lunch on a table someone had set up, either the Red Cross or the Salvation Army—and boy, we were hungry!"

The men were working so hard, Cecil claimed, that a break was required due to the fact that they were simply famished. Food, he suggested (and suggested only because he didn't remember clearly, thus couldn't say for certain) was provided by the same charity organization that set up the tables.

But Cecil did remember one point, one incident of many like it that continued to play out over and over anywhere the men went to load up the trucks with all the trash and rubble that the tornado had created from lives and livelihood. It was the only thing Cecil conveyed that indicated the severity of the trauma on a more human level than Model T dump trucks full of trash.

"There were people walking around looking at their homes, crying," he said, again in a rather flat, emotionless tone, as if it were safer than the alternative. Cecil seemed to put forth an effort to distance himself from the horror as he admitted, "There was a lot of damage."

Upon return to his downtown Zeigler apartment, Cecil met up with his roommate, the equally exhausted Dr. Tashma.

"The doctor told me *nothing*," Cecil proclaimed, shaking his head in slight disbelief. "Nothing but that there was a lot of damage and he had treated a lot of people for cuts, bruises, and broken bones."

Cecil smiled a faint smile. It was as if he were fully aware that those minor-to medium-severity wounds were not all that the good doctor saw in his first night of practice following the tornado.

"He was very closed-mouthed about the whole thing and his part in it. I heard a lot of people talk about it six to eight months or a year later; they'd tell about their houses, people being injured, planks being blown through trees…

"The fact remains that if there was anyone else with trucks over there in Murphysboro, I don't know," Cecil went back to his original involvement in the hours following the storm. "But we did our full day cleaning up the town of Murphysboro. The trucks held not as much as the big trucks; they were just Model Ts, but the beds of the trucks were always as full as we could get 'em and dump the stuff and go back.

"Many little things happened when we were there, things I don't remember all of. But I do remember people would wave as we went by, saying 'Thank you very much, we appreciate it. Thank you, folks.'…"

It was at that point that Cecil Spain's voice finally let go with some odd, full emotion instead of the professional tone that he had, up until now, successfully kept up.

"It never dawned on me that we were in such a trauma as a result of this 'windstorm' that hit Plumfield and missed us in Zeigler."

* * *

Plumfield, Illinois, had a green and white sign directing the driver to its basic location. But aside from paved streets laid out in a grid and very neat, often very pretty and expensive-looking houses, there was not a lot else there. It was not so much of a town or a village than it was a settlement—no businesses, just residential.

Located on the other side of the Big Muddy River from Zeigler, Plumfield, it was said, was denied protection from the elements by the river in the same manner Zeigler *was* protected by it. It was an old superstitious tale in the area, but it seemed to have its basis, given that the death toll from the tornado came directly from the little settlement—24 people in the 'Zeigler Area' of Plumfield lost their lives on March 18, 1925.

Mary Parrish, 73, of Plumfield, knew well enough about the tornado even though her birth took place two years later. Her parents, Harvey and Lillian Wallace, had related the tale to their daughter over the years.

"Mom and Dad had just built a new home," she said, "and it withstood the storm, but the garage went.

"Further down into Plumfield, it took the school and killed two little girls there," Mary, who lives beside the river, said. "The tornado went a little south of where my house is now, and Mom and Dad always said that people were blown all over from inside their homes. At Wesley Gunter's, a baby was pulled out of its mother's arms and was found dead in a field.

"Then it went on in to West Frankfort," she said simply.

"It came from out of the blue," she said of what her parents always told her. "Dad had just put the car in the garage and the wind came up and took the garage away.

"My aunt and uncle, Norma and Elmer Wallace, were going into West Frankfort, and they saw it coming and got out of their car and into a ditch."

That move, noted Mary, saved their lives.

The Big Muddy River ran near Mary Parrish's home, which was not far from the original home place where her parents were living in 1925. Mary had this tale of what they experienced when the storm blasted through little Plumfield.

"It followed the Big Muddy from Murphysboro, and the river runs out here by our place," she said. "There was a big bridge with a metal top on it near the house, and Mom and Dad always talked about how they could hear that metal rattling when the storm went over."

*   *   *

The thought of the sound of metal rattling in winds exceeding 280 miles per hour, sustained at a forward moving speed of 56 miles per hour, was a haunting and unnerving one. It was likely one that stayed with Harvey and Lillian Wallace until their dying days.

The interior of the community storm cellar at Plumfield.

The terror of a tornado of this magnitude and proportion left an indelible mark on the minds of the residents of the little settlement as well. In the years following the tornado, the specter of the winds remained with those who could ill afford to lose everything they had to them, but could also ill afford the means by which they could at least escape with their lives—via storm cellars. So a community storm cellar was installed by professionals who scoped out the areas hit by the 1925 twister, and capitalized upon the possibility that the fears of their residents would prompt them to take precautionary measures for the future.

The storm cellar sat embedded in a lot next to the Plumfield Christian Church, which was a north turn off Highway 149 and the entrance into the Plumfield community. The cellar was of heavy concrete, vented at the top by four smooth brown concrete air pipes, looking more like small chimneys than vents. It was about twelve feet wide by maybe thirty feet long, perhaps eight feet tall, and very sturdy. But in 1999 it was little used and not well maintained. Six steps lead down into the chamber, but they were overgrown with weeds and covered with leaves, and the interior appeared as though it hadn't been cleaned out for quite awhile from the normal debris and water accumulation and plant and animal life that can leave their mark over several years' time.

Notably, in later years and just prior to this book's publication, more care had been given to the shelter, likely from the nearby church, but one couldn't help but wonder: How often was the cellar used? It was in an odd location if it was; on the outer edge of the community, and in a storm the size of the one for which the cellar was built to preserve life, the likelihood of its effectiveness was good only in the case of those in closest proximity.

Some may have looked upon it as an eyesore, and wondered when the community storm cellar would begin to crumble and collapse, and have to be excavated before it became a hazard to the neighborhood children. Many of those same ones wondered if a storm of the magnitude that prompted its construction would ever bear down upon the area again—and not when.

And still others, senior citizens who happened to be going past the structure infrequently in their day-to-day existence, looked upon the old storm cellar as a symbol of comfort, and of refuge. It stood as a barrier against the elements, if not against time, the latter of which could not erase the memory of that afternoon in mid-March 1925. Regardless of whether the storm cellar had ever or would ever be used by them, it stood in wait.

It was a lonely sight, and a reminder, to those who knew why it existed, of an afternoon many years ago, of just what nature could do when the skies turn black, the winds turn deadly and wicked weather turns its eye toward little towns like Plumfield, Illinois.

Chapter Ten

West Frankfort, Illinois

The pride of Franklin County, for many, many long years, was believed to be West Frankfort. Some considered it to be a dirty, gritty, mining town, not worth the attention it was shown due to being one of the greatest coal-producing areas in the state. At one time, West Frankfort had some of the largest coal mines in the world. It not only carried this distinction, but it had a very active railroad hub, much like Gorham and Murphysboro had had, only not quite on that scale.

But West Frankfort had its drawbacks, too. An active Ku Klux Klan membership, burgeoning up from troubled Williamson County to the south, had a tight grip on the community in the 20s. Rallies and 'meetings' stirred the town and its residents and created an aura of fear within the city limits as well as without. Crosses burned frequently on lawns of those who met with the Klan's disapproval. While Williamson County as a whole became known in the 20s as 'Bloody Williamson' due in part to the Klan's activity as well as the miner's union strikes and the difficulties involving racial prejudices, Franklin County did not escape ignominy. In fact, it was widely known that when the tornado made its terrifying way into West Frankfort, a Klan rally was being planned for an area outside West Frankfort on that very night. Many were already shuddering in fear of which way the winds of the powerful organization were going to blow next. They could not foresee a literal wind of devastation beyond what was about to take place among their very own citizens.

Upon our first visit in July 1999, my children and I discovered West Frankfort to be misleading to the uninitiated as far as its layout went. Off Interstate 57, Highway 149 had taken us to the smaller towns west where the tornado had struck. We had initially gone around West Frankfort that first night, intent upon examining it on the way back through. We knew we were at the western edge of the town as we rolled through. We just didn't know how far east to west the town actually stretched. We found out on the return visit.

East through town, West Frankfort's business sector went on for blocks and blocks and blocks. Past business after business, I was heard to exclaim many times, "How long is this town, anyway?! Good grief!!!"

Finally, we emerged on the east end, where signs told us that a coal mine museum, which actually took the visitor underground to a real mine, was open to the public for most of the year. The National Coal Museum went to a depth of 600 feet and acquainted the visitor with the innerworkings of a real mine. We did not have time for a tour, but put it on the list of things to do once the research was completed.

Another fascinating sight, and also one of those things on the list, was the Franklin Area Historical Museum, located in the old Logan School building on a hill on the northeast side of the town. As in Murphysboro, DeSoto, and points in between DeSoto and West Frankfort but further to the south, it seemed the recurrence of the name "Logan" was an ongoing theme. The museum kept regular hours, and was filled top to bottom with artifacts of an earlier area, carefully collected and displayed with the concern that only those with a deep appreciation of history and what can be learned from it can possess. The area in which the building sat was known as Tower Heights. Long ago, in the early 1800s, there was a palisaded fort built on the top of the hill as a defense post; many of the sort were built along a trail through Southern Illinois from Equality toward Kaskaskia. Constructed by a family whose apparent leader was Francis Jordan, the area was later slurred and slanged and finally came to be called "Frankfort" being Frank's Fort, after all. The fort came down, newer structures came up, and the water tower was built on the hill where the fort that was Frank's originated.

There was a burial ground toward the back of the museum called Tower Heights Cemetery, and it stretched across rolling hills with a beautiful and peaceful view. We had no success finding gravestones marked with the date that we sought within that place of rest.

However, at a little restaurant on the west side of the city, where they served the best blueberry pancakes my little daughter and I ever had the pleasure to enjoy, we did happen upon one survivor with an incredible story to tell. Hungry's, as the restaurant was known at the time, was constantly a busy place; we ended up visiting it time and again even after the collection of stories we acquired there were complete in 2000. But never would we hear tales more riveting within the city of West Frankfort than the ones we got out of Hungry's that first time. Their customers were young and old, visitors and residents. But the look on the faces of those who were survivors told me all I needed to know—

That just the mention of the 1925 tornado was enough to stir the soul as well as the memories of those asked.

"I opened the door and saw a whirlwind of glass..."—Arthur Reed, age 9 on March 18, 1925

Arthur Reed could say with fairness that he owes his existence to a toothache.

It's one of those funny things a person could look back on and laugh about, almost, if one could laugh in the face of such a tragedy as that which befell the city of West Frankfort on March 18, 1925. The average nine-year-old boy, prone to play hooky, especially if the weather was nice, was going to take advantage of an opportunity to stay home from school—genuine toothache or no.

But that was what kept Arthur home on that day, and had it not, the story he told might not have had the outcome it did—and as it was in DeSoto, though not to that hellish scale, the schoolhouse in West Frankfort was not the place to be on that dark, warm and cloudy pre-spring afternoon.

A lifelong resident of West Frankfort, Arthur Reed enjoyed his coffee in Hungry's, West Frankfort's premiere pancake restaurant in late 1999, and told the tale of what happened on that long-ago March day when the winds grew wild and the sky turned to pitch.

"I was home from school with a toothache," Arthur grinned with perfectly good teeth, "and I was with my sister and a baby nephew on our back porch cracking hickory nuts to make chocolate fudge."

At the sound of his own words, the contradiction of which was made in one breath, Arthur sat back with his coffee and laughed considerably, as if to admit, 'okay, maybe the reason why I had a bad tooth was because of the hickory nuts and the fudge.' But he said nothing further on the subject.

It had been a lovely day, warm, almost muggy, with only a few dark clouds in the sky. The heat rose only a couple of degrees by afternoon, and Arthur likely, by that point, had not regretted his decision to stay home and nurse the tooth. His sister, Faye, and nephew R.H. Mosley were as startled as Arthur was when mom Anna Reed suddenly came to the door with panic edging her words.

"We were on the east side of the house, but I could see around the corner that the sky was black and green. And then I heard Mom say, 'You kids get in the house; it's gonna storm!'

"And boy, did it!"

Down the middle of their modest house, the Reeds had a sort of long hallway that was doubling as a closet.

"It had hangers in it, and I remember as we were going in that the hangers were rattling as the house was being shaken.

"My big sister was in the back bedroom with my nephew and my baby sister, Peggy Ann, who was just eight weeks old, but I didn't want to stay in there with them. So I went down the hallway and started to go into the dining room," he said.

And then his voice got quiet, not so amused any longer that he was staying home with a toothache on that very day.

"I opened the door and saw a whirlwind of glass, so I went on to the back bedroom." When asked to explain what he meant by a 'whirlwind of glass,' Arthur only shook his head and shrugged. His best guess, he claimed, was that the vortex was likely just above the house or creating such a vacuum within it that when the dining room windows were shattered at the approach of the storm, the shards were forced to the center of the room and suspended in the vortex instead of being sucked out. At any rate, the sight of such a deadly, shredding glass tornado within his own home was almost more than little Arthur Reed could take. He backed out of the room and headed to safe haven, or at least, what most of his family thought would be safe, in the back bedroom.

> *"My older brother Lawrence told our mother he heard something in the living room heating stove telling him not to go to school"—George Gossett, age 5 on March 18, 1925*

Gayla, a waitress at Hungry's in West Frankfort, knew of another 1925 tornado survivor and was happy to give out the name and address of the gentleman whom most of the town, she assured, knew as such.

In late 1999, George Gossett and his wife lived in a pleasant house on

Stunned West Frankfort residents assess the damage in their town.

East Oak Street in the middle of West Frankfort with their little dog, Chico, who had an unusual but delightful talent—

Chico would 'sing' when George played his harmonica.

And George was a very talented harmonica player. He passed the time that way, growing weary with the years upon him, he claimed, although he had a grin and a demeanor of a man very much younger than 80 years.

Not long ago, he revealed, his daughter, who was a schoolteacher in Carterville near Marion, called him and requested that he write down his memories of the day that the tornado of 1925 took away at least twenty percent of the buildings in West Frankfort, injured 410, and killed 148—the second highest death toll per city the tornado claimed, right after Murphysboro. George's daughter wanted her father to recall the incident for her class as a sort of study unit—and George Gossett responded by filling up eight handwritten pages with his own memories of what took place on that dark and deadly muggy March afternoon.

"The Great American Tornado of March 18, 1925" was what George called his composition, and as he referred to it during the telling of his personal account. "That's what at least one encyclopedia calls it."

"Being short 41 days of my fifth birthday (back then) and am now over a half year past my 79th, I still remember that day fairly well," George began by reading his own prose.

A West Frankfort man sits among the ruins of his home and neighborhood.

He went on to describe the morning as having "blustery, small, dark clouds passing swiftly." Elaborating on his composition, he said the air felt 'close' and rather uncomfortable, and had all day long.

George had four brothers at home with him that day, John Bowen, almost 13, Lawrence, 8, Chester, 3-and-a-half, and Roy, almost seven months old. It was Lawrence who had, earlier that day, opted to stay home from school, explaining himself in that eerie way only children can, sometimes being open to sensory perceptions in a way most adults cannot.

"Lawrence told our mother he heard something in the living room heating stove tell him not to go to school," George said, "and she, being of the superstitious type, kept him and our oldest brother, John Bowen, home."

Their mother, Winnie, took the littlest brother Ray into town, pushing him in a baby buggy.

"I remember seeing them coming home, probably a couple of hours before the tornado struck," George said.

"She said she saw a huge black cloud coming out of the west by southwest, like it was right on the ground," George claimed. "She had John go to the well off the back porch and draw a bucket of water, and throw it in the heating stove to extinguish the fire."

This was the very same stove that, earlier in the day, Lawrence had heard tell him not to go to school. It was shortly after three in the afternoon. Most of the children were only just leaving the school grounds; the Gossett children were not, thanks to Lawrence, among them.

Throwing the bucket of water on the coal burning fire in the heating stove cracked the stove, the shock was so sudden, reported George. That was the last thing he recalled before the next instance, when his mother moved to get them into the northeast bedroom.

"There was a sharp streak of lightning and a loud clap of thunder, and windows started breaking," George said. "Mom got us into the bedroom and somehow blocked the door and it stayed shut.

"There was one window to the east and one to the north in the room, and I looked across a small table through the north window, which never broke. At first I could see the house across the yard there and some young maple trees between our houses.

"Then, it became so dark I couldn't see either, the house or the trees."

*   *   *

After his experience with the whirlwind of glass, Arthur Reed had decided it would be best if he got away from that room, the dining room, as quickly as possible.

"When I went back to the bedroom," Arthur said, "I reached for the hall door and that's the last I remember of that.

"Something hit my head, probably a nail from a board…the house was tearing apart…and this, whatever hit me, fractured my skull. But I kept going.

"My sister Faye, who was 14 at the time and in the back bedroom, said when I backed up from the doorway, I took hold of the foot of the iron bed where she was and me and the bed were flying through the air.

"That storm let me down three lots from our house on North Illinois Avenue."

*   *   *

Again and again, in the tales from Missouri and thus far throughout Illinois, George Gossett's next feature of the story was repeated; and, as incredible as it sounded, it never lost its impact but that impact always seemed to *gain* upon hearing it repeated.

"Our house started to rise *three times*," George insisted, although, he said,

"I didn't count the times, but my mother and two older brothers did.

"One.

"Two.

"Three.

"When it came back down, it twisted and crumbled the foundation. And when the tornado passed us, the house to the south was completely gone and the trees were in bad shape, although it seems at least one of the trees was still living when I passed there recently.

"The house west across the alley was gone. Next door south, the whole roof had blown off, the same as the house across the street southeast."

George went on to say, "When we came out of the bedroom into the living room, I remember seeing chicken feathers and bricks in the floor."

*   *   *

When Arthur Reed regained consciousness, there were two or three men trying to lift a section of the Reeds' house off his mother.

Bringing himself around, Arthur got up and went to help lift the section.

"But when I bent down, I couldn't get back up. They thought my back might have been broken, but it was just a strain. Still, they strapped me to a board and a delivery van came and took me to the hospital on St. Louis Street. The hospital was full!" Arthur exclaimed. "And they wouldn't accept me."

Apparently, in accordance with the injuries the medical community of West Frankfort was seeing, Arthur Reed's injuries, whatever they were, were minor; he was conscious, not bleeding severely, and had all his limbs despite the skull fracture he'd sustained. Conversely, the more severely injured were lining every area of the medical facility, hoping to get to a doctor or nurse for treatment before their condition worsened.

"Three flights of stairs up to each entrance," claimed Arthur, were crowded with such grisly conditions.

"They took me uptown," Arthur said, "and that was a hard thing to do, there was so much trash all over the place. It took thirty to sixty minutes. And they dropped me off at the Masonic Hall, where they were taking care of people who weren't hurt too bad."

Back at the home place, it had happened that the winds had blown the walls out and the roof off; in Arthur's estimation, those who managed to remain inside weren't much in harm's way.

It was the ones who were blown outside, as Arthur and his baby sister were, who suffered the most.

"About Peggy Ann," Arthur reveals of the eight-week-old infant girl, "during the storm, Mom went to pick her up out of her crib and that's all she remembers.

"They found Peggy Ann under the same piece of house as Mom was, a long section, probably twenty feet. Peggy Ann had drowned in a puddle. There was not a mark on her except for a scratch on her nose and some bruises."

Arthur said about his other sister, and little nephew R.H., "Faye was tending to the baby boy after the storm; they didn't have any storm-caused injuries until after she went to walk around and got into a high-tension wire. It caught her on the back of her knees, and it burned both of them, because she was holding R.H.

"There was just all this rubble everywhere and she was trying to avoid it."

*   *   *

"My family didn't receive a scratch," said George Gossett.

"After it was over my two older brothers, John and Lawrence, jumped out the living room windows, but our mother wouldn't let my younger brother or me jump out," he remembered, and from his statement one could almost hear the disappointment in his voice over the fact that George and his younger brother were 'too little' for Mom to let them go jumping out the window after the storm.

"We couldn't get out the front door, as it was jammed shut, until a policeman came and knocked it open with either a sledge or a brick."

George went on to say "Most streets in town were still undriveable, and people were carrying their injured toward the hospital using whole beds for stretchers. They weren't about to walk a straight path because of debris in the streets, like furniture, pieces of cars, wagons and buggies, and just about everything else."

*   *   *

Clyde Reed, Arthur's father, was in the C.W. Neff Coal Orient No. 2 mine when the storm struck, and had come walking back into West Frankfort to search for his family. The coal company had brought the men up out of the mines, told them there had been a storm, and had let them walk home. By the time he had arrived, his family had been taken to other locations, and the long section of house that had trapped Anna Reed had been removed.

This may have been the very same well curb that Clyde Reed found, all that was left of his family home, when he arrived back in West Frankfort on March 18, 1925, after working in the nearby mines.

"The only thing left on the lot where our house had been was a brick well curb, and this had a walkway out to it.

"Dad had an aluminum dinner bucket that he'd always carried his lunch in to work. And when he got to our lot, he saw that all that was left of it was the walkway up to what had been the well curb, and even part of that was gone.

"Well, he walked around that open well two or three times, and finally, he said, 'If this is all I've got left—!!!' about his pail, and he threw it into the well. He didn't want that pail if that was all he had."

Eventually, the Reed family gathered back together, and there was a brief funeral service and interment of the baby Peggy Ann. However, it took Arthur Reed's father Clyde Reed three days to find him after the storm.

"Dad and my brother Donald, who was 16, was looking all over town. They were looking at death lists and at injured lists. They came across the list at the Masonic Building where it had my name and another name; they didn't know which one was alive and which was dead so they decided to check it out.

"They found me in the middle of the bed playing marbles!" Arthur laughed. "My brother still kids me about being there; they could hear me hollering 'anys' and 'nothins' when I was playing marbles. I'll never forget it!"

The stay wasn't the most unpleasant thing that had happened to Arthur, to hear him tell it.

The storm toppled the water tower at the Orient mine.

"There had been blood all over me when they brought me in, and the nurses had cleaned me up and took care of me. They'd had to shave my head where the nail hit me, and bandage me up good, but they kept telling me 'Your family will be along pretty soon to get you,' and I believed them, otherwise, I'd have been pretty upset."

Anna Reed, however, did not fare so well.

"Mom wasn't hospitalized, but she was 'addled,' grieving. I didn't even get to go to the baby's funeral," Arthur noted sadly, "because I was in the Masonic Hall."

It took his family, and in particular, his mother, a very long time to get over the death of their precious baby Peggy Ann. But in the days that followed the storm, they had enough to contend with in the physical manifestations of the storm's presence long after it had gone.

"On me, my mother and my sister," Arthur said, "little pieces of sooty stuff from the flue were driven into our skin."

*　　　*　　　*

West Frankfort was just up the road from the practically new Orient No. 2 coal mine tipple, which was known as the largest mine in the world for breaking the eight-hour world hoisting record. The storm, a mile wide in diameter when crossing over the northwest part of West Frankfort, left the

Caldwell Mine No. 18's battered coal tipple crushed by the powerful winds.

Orient No. 2 mine tipple stripped of its sheathing, exposing the battered skeleton. But it was a far cry from being the twisted, useless piece of wreckage that the coal tipple at Caldwell Mine No. 18, on the eastern edge of West Frankfort, became in the winds, and it was all the more horrific that it was made that way in a matter of minutes.

In the town and outlying areas of West Frankfort, 148 people died, a startling 410 were injured, and $800,000 in 1925 dollars in damage was sustained in the wake of the monster. And still it did not lift from the ground, but instead drove with insane and insatiable devastation across the farmlands and above the coal mines of south central Illinois. West Frankfort experienced particular damage due to the fact that the tornado, which had slowed to its slowest forward speed upon leaving the Bush/Hurst area, was moving across the land at a mere 56 miles per hour. Taking its time now, it still moved at twice the rate an average tornado travels. Its winds were estimated at its developed standard of 225-280 miles per hour. The odd, 'three-gust' sensation that witnesses reported from its origin in Annapolis, Missouri, continued.

Twenty percent of West Frankfort, the northwest side, lay in ruins. The massive twister, if indeed it were only one, had claimed the lives of 541 people and had left 1,423 seriously injured in the span of only 40 minutes from leaving Gorham to attacking West Frankfort.

But notably, West Frankfort was the halfway point of the storm. It would continue on its N 69° E heading for the rest of its trip across the state of Illinois, and beyond.

Ruins of the C. W. & F. Mine in West Frankfort showing the main mine tipple (top), the shops (middle), and the damaged cars of the miners in the parking lot (bottom).

Chapter Eleven

Parrish, Illinois

On a warm autumn afternoon in early November 1999, my 3-year-old daughter and I took off for one of the communities struck by the 1925 tornado closest to our home and within a reasonable hour's driving distance. It was Parrish, Illinois.

Parrish had the distinction of being the last viable community listed on the Illinois roster of those damaged or destroyed by the killer tornado. After Parrish, the storm skirted several other active communities, passing either north or south of them, but did not incur enough damage within any city or village limits to be able to lay claim to strikes within towns. Parrish led the next entry into the damage rolls as it listed 'Rural area—Parrish to Crossville.'

There was no difficulty in getting there; a trip via Interstate 57 from either the north or south to Benton, a sizable community, would take the visitor to Illinois Highway 34. From there it was a heading of southeast for only a few miles to the road that lead to the area known as Parrish. Little communities, which owed their existence to the mining industry, were one after the other on the highway that ran from Benton to Harrisburg. With curious little names that reflected either John A. Logan's ongoing influence in the area or the direct and more potent influence of mine works, yellow diamond-shaped signs alerted the driver that there was a turn off the highway ahead for each name.

A short blacktop road past a couple of fields, and there was Parrish.

One would likely never know, were there not signs off the highway to indicate the community even had a name. By 1999, a few pretty, new houses stood in stark contrast to several ramshackle dwellings and rundown outbuildings; some of the latter appeared to be on their last timber, as if a sigh of wind may very well end their stand against time and the elements with a shuddering last gasp and collapse of beams. A Baptist church rose proud to the west, the size of it bearing out that likely it was a one room sanctuary which, on a good day, might have held 50 or so parishioners and no more unless they were standing behind the back pews.

Further down the road to the south, past several more rundown homes and a shotgun building coated in blood-red paint and trimmed in white (and where gloriously-colored roosters and hens darted about, crowing and clucking), some newer houses and mobile homes finally ran out of street. The road abruptly ended in a T. To the right, fields opened up and were embraced at their periphery by heavily wooded areas; to the left, Parrish continued. Lots that may once have held more homes held only prairie grass which grew unchecked and untended in the early days of fall. Ahead, a gravel road veered to the right at about a blocks' length. A defunct Methodist church sat at the juncture of gravel and pavement. It silently bemoaned the fate of unattending Methodists. Parrish continued for a little way, maybe another half block, to the left.

The street, if it could be called such, had the distinction of encircling the block where the only visible sign of a rail line could be discerned within the town, unless one was otherwise looking closely. Behind small homes and more outbuildings was a neat little rise, perfectly symmetrical, a long, grass-covered, snake-like mound that bisected the entire block and mysteriously disappeared as it met the blacktop of the street on the north end of the block. The mound remained hidden in the prairie grass...and reappeared on the other side only briefly after it crossed the street upon which one first entered the village. Here, the ghost rail line reemerged and wended its way into a grove of trees. It was easy to imagine that at one time, those trees that formed a perfect canopy above the ghost mound were mere saplings. The rush of a train zooming by could have sent the saplings bending, and half a century ago, there was no doubt that it did just that, for the train did not stop in Parrish.

Not like it did in the days before Wednesday, March 18, 1925.

> *"After they thought they'd got all the bodies... farmers kept finding more in the fields"—Lena Ross, age 8 on March 18, 1925*

"There were a lot of fine houses in Parrish back then," Lena Ross said of her little hometown in the 1920s. "Everybody worked in the mines and made good money."

The village of Parrish was a viable town back in the days of the mid 20s. Counting those on the outlying farmlands surrounding the village, there were almost 300 souls who called Parrish home. Times weren't necessarily easy, but folks were able to make a living farming or working in the mines, sometimes a little of both. Farmers were a hardy and relatively successful breed in the area of Parrish, even given the depressed economic conditions of farming following World War I. The land existed on a former floodplain, almost a valley of sorts that ran north and south; massive floods were infrequent in later years, but long ago, they deposited rich topsoil in this stretch of Illinois. Those in Parrish farmed the land to abundance; animals of all sorts also inhabited the community, and pleasant mooing, whinnying, baaing and clucking would accompany the sounds of day-to-day community life here in the early part of the twentieth century.

As in many southwestern Illinois areas, coal was discovered to be plentiful in that part of Franklin County, and was mined from earliest settlers' times. Back then there were many mines to be found in the region. Illinois Highway 34, as it came to be known, was littered with them in the 1920s. Northwest of the area was Sesser; the Orient and Pershing mines were directly west; McGee Mine was located in Galatia southeast, and the Logan mine was directly northwest of Parrish by only a few miles in those decades ago.

Lena Ross was a well-known Parrish resident; ask anyone in the village, and they could to point you in the direction of her modest little home in the middle of town. Oddly, it sat with the snakelike mound of the former railroad track crossing her backyard.

"We had chickens back there for a long time," she said, "and they were ready for the trains and knew when they were coming through, so they'd run up next to the house at that end of the pens, and tuck their heads and be still when the trains came in."

Her family, she insisted, got used to the rumble of the trains as they whisked past her home.

Her home had a few signs of age on the face of it; the concrete steps were cracked and a little crumbly as they made their way up the front porch, and her son John warned anyone traversing them to "watch your step, there!"

On a quite pleasant fall afternoon, a week past the time change from Daylight Savings to Standard, the sunlight reached in from the west. It seemed to embrace Lena as she rested in her easy chair, a once-modern recliner of nondescript beige. Little tufts of white fluff now stuck up where the edges had been worn at the ends of the armrest; just a few cracks of the vinyl upholstery could be seen at these points. The rest of the chair's vinyl was worn smooth and a little dull with use and age. But the vinyl creaked when Lena sat a bit forward as we entered; the old chair had probably not even seen its best, most comfortable days just yet.

Lena's white hair was thin, her face webbed with fine lines of wisdom, but her eyes twinkled and her voice carried a lilt in the greeting she gave, whether she wished for that lilt or not.

She was a thoughtful woman, speaking only after reflection or consideration. Initially, she didn't want to speak at all of the weather disaster that changed the lives of everyone she knew and devastated her hometown.

"I lost so many friends and people I knew…;" she offered at first, hoping to explain her reluctance.

Her eyes averted, her head lowered a bit, and the web of facial lines blurred softly, away from the sunlight. It was very obviously a trauma still relived by this woman after so many years.

But when asked about her home town, or the vestiges thereof and of what used to be, Lena Ross was suddenly more animated, willing to voice her opinion, and it was apparent this was a subject dear to her heart.

"It was a good old town! Pretty. Pretty farms on the outside all around the edge," she blossomed.

"Miners made good money," she pointed out, "and lots of them owned homes right here. You wouldn't know it by looking at it now," she said, her voice more serious, "but there used to be two or three grocery stores here. There was a clothing store, a filling station, a post office," her eyes gleamed, "and a train depot.

"We had all that until the storm came."

And now Lena Ross was ready to talk about it.

"We came back to the house and it was gone. So we thought, 'Oh, everybody's dead'"—Adrian Dillon, age 12 on March 18, 1925

Adrian Dillon, 87 years of age around the date of the 75th anniversary of the great tornado (March 2000), lived in Waltonville with his wife Wiladene in a restored bank building in the center part of the town, which was to the south and west of Mt. Vernon, Illinois by thirteen miles. The house was beautifully decorated with articles and items from Adrian's time as a merchant marine in World War Two. Fans, cloisonné enamelwork items and ivory carvings all adorned the walls and floor spaces of the former bank building, but many of the standout decorating items were clocks.

There were flat clocks sitting on coffee tables, face up; one mantle-type clock squatting, poised, on the hutch; two Black Forest-type cuckoo clocks stood at attention against facing walls; and one clock that was cloisonné hung on the wall and coordinated with the other enamelwork items hanging, surrounding it.

The cuckoo clocks chimed in succession.

"Sure do have a lot of clocks," he commented, seated in his easy chair with his walker in front of him and his oxygen nearby.

The expression on his face, which was weathered gently with time and the many years as a member of the Armed Forces and as a successful business owner, explained his next statement. He knew full well that all the cuckoo and cloisonné and fat wooden mantle clocks would only tell time, not slow it down while counting it.

"We don't need 'em, do we?" he asked rhetorically.

Adrian's memory of the tornado had inspired him to hang on to newspaper clippings and information about it, which he kept in a file folder in a box tucked away within the recesses of the bank's vault. The room, the entryway of which was still marked with 'Waltonville Bank' above the heavy vault door, was where the Dillons kept their memorabilia. It wasn't just of the 1925 tornado, either; other storms that had struck the Franklin/Hamilton/Jefferson County areas were stashed away, each account within their own manila folder or envelope. Adrian wasn't obsessed, but he was knowledgeable, and especially about the 1925 storm, of which he was a survivor.

"I got out of school that day to play in a marble tournament," he recalled with a certain hint of fondness in his voice over the memory. He continued that line of thought. "I was nearly 12 years old, sixth or seventh grade. They had tournaments. You know how now they have basketball and football? Well, back then it was marbles. There were several in it; they'd hold it around the depot and post office area, where the trains came in. They had chat rock laid down, and that was where we'd make all our rings."

The day was pleasant, Adrian related, warm and spring-like. He, and the boys with him for the tournament, gathered at the depot area, not at all suspecting that their warm, pleasant afternoon out of school was about to turn as chaotic as they had ever known.

*   *   *

"It was a hot day for March, that's for sure," Lena Ross recalled in a musing way. "We had the windows open in the school that day."

The school was called simply Parrish School, of the one-room brand of schoolhouses used back then that held all levels of grades and ages, and employed one schoolteacher to deliver all the lessons in segmented time frames.

The man of learning, Lena pointed out, was Willard Carlyle, and he was of such a presence of mind that he fidgeted in the warm March afternoon. It was, Lena almost alluded, as if he were able to sense the pending arrival of the monster that had already decimated parts of southeastern Missouri and Southern Illinois in its path.

"He was nervous," she said simply. "He had us take up our books, and acted like there was nothing wrong.

"But then that big black cloud came out of the southwest; we could see it coming." Lena touched her fingertips to her lips and paused in this way, thinking. "It came up so fast, I remember that."

> ### *"(Mom) had a kid under each arm when that storm hit"—Wendell Dillon, age 5 on March 18, 1925*

Wendell Dillon, who was 79 in late 1999, lived in West Frankfort, Ilinois. A brother of Adrian, he was but a babe of five in Parrish when the monster in the clouds emerged from the southwest; he and his sister Faye, just a few years older, were with their mother on their farm located about one mile northeast of Parrish. His version of the story enhanced Adrian's, and, as related to him over the years by his mother, it was a very straightforward one, and one he shared in the telling with his nephew, Marv Wilson.

Marv continued to reside in Parrish with his wife, Carol, on the location where Marv's father, Ed Wilson, had kept a homestead many years after the killer storm finished its gruesome work. On the farm stood a barn; incredibly, it was one of the very few buildings that withstood the terror of the winds as they screamed across the minimal acreage of the little village. The barn had been shored up over the years, and stood in defiance of the elements.

It was as if it watched as Marv assisted with Uncle Wendell's tale:

"Grandma always told me that she could see a cloud coming up late in the afternoon that day," Marv related, "and watched it getting worse and worse and decided something was going to come out of it, so she took Mom (Faye Dillon Wilson) and uncle Wendell towards the front of the house."

"She had a kid under each arm when that storm hit," laughed Wendell of his mother.

Their mother was no doubt at a complete loss with what to do and likely in an utter panic.

"Dad was in Parrish, so we were at the house by ourselves," Wendell reiterated, not at all disturbed by the memory of the moment when the killer storm began to claim its next set of victims.

*   *   *

Adrian Dillon, who was on the south end of town from his siblings and mother, remembered in a bit more detail than his younger brother Wendell.

He recalled looking toward the west where sat, of all places, the Ross farm where Lena and her family resided.

"It was about a mile from where we were," he said, "and all of a sudden it got really black over the horizon.

"I seen that tornado pick up that Ross barn and house and buildings…" Adrian remembered, his words trailing slightly. "And when the tornado headed into town, me and the four other boys ran into a building; I think it was the Crisp Grocery Store."

Adrian, visibly upset, continued despite the terror of the memory.

"It was a loud roar," he admitted shakily; "a loud noise. A very *wide* tornado. I saw a giant funnel come across that farm, pick up a building, and blow it away."

Adrian looked up from his reminiscing and pushed further into memories of that dark day.

"It was a little wider than most tornadoes," he said. "For a quarter mile, in the middle part, it ground everything up. Then, that much on either side of it blew up houses, barns, *everything*. Just blew it up. So I would guess by that it was about three-quarter miles wide."

Something about the menacing appearance of the approaching storm, more than just the sound, the width, or the obvious destructive nature of it as it tossed entire structures about like stacks of twigs, prompted a definite response in the 12-year-old Adrian.

"I said, 'We have to get back to the school,'" he reported telling the other boys with him at the Crisp Store.

Apparently the boys did not argue. All five began a mad dash down the railroad tracks, quickly crossing, with the speed that only young boys in a panic could manage, the few blocks' space between the store and the Parrish School.

"And as we got back to school," Adrian said, "the high winds were blowing out the windows. We could just hear them crashing.

"But we ran inside anyway."

*   *   *

"It got *close*," Lena Ross said with obvious emphasis on the word 'close.'

If there were other words or phrases to describe the feeling of the massive cyclone as it passed over Parrish School, Lena didn't use them nor acknowledge them. Rather, her arms crawled from beneath the afghan she used on late afternoons and early evenings in the fall and winter. She raised them high in the air. Despite her apparent limitations due to age and the accompanying ailments it tends to bring on, she spread her hands wide and brought them down to waist level, and back up! and back up! and back up! as she explained:

"It closed in three times.

"Whoosh!

"Whoosh!

"Whoosh!

"And the third time, it was over.

"It blew all the windows out," she remarked, again, a simple observation that needed no elaboration.

"The windows were set in squares," Lena described. "And there was a kid, 'Pancake' Patterson, that the big squares hit. He didn't get hurt," she said with wonder. "But he was scared."

*   *   *

Wendell's and Adrian's father, John Dillon, had been in the village of Parrish for some time and was standing between two other men from town, chatting in one of the grocery stores, another shotgun building located next to the railroad depot and tracks.

The day's-end chatter was common about this time of day year-round except during planting and harvesting. Planting season was a mere couple of weeks away. This was likely what the topic of conversation was in the little village store when the storm swept down from dark skies on them all.

"It blew the store away," said Wendell Dillon.

Everyone inside the store was swept up by vortex winds that were later estimated to have been close to, if not in excess of, 300 miles per hour, this estimate made by the surveyors' findings as they were reported to weather experts of the day. Winds that had rotated rail cars and axles at locations across the damage path now swept through smaller and lighter venues: furnishings and fixtures inside the store in Parrish were carried away like feathers in a breeze; the men inside—the same.

"Dad came to by the railroad tracks," said Wendell. "The other two men were beside him.

"They were dead."

*   *   *

And the storm had passed the schoolhouse of Parrish.

As if the Parrish School teacher, Willard Carlyle, wasn't nervous enough before, now, Lena Ross pointed out, he was at his wit's end.

"I've never seen a teacher so nervous," she said with some wonder.

"But the school house was still standing, and none of us were hurt bad."

By large contrast, the Parrish School building was a safe place to be on that dark March afternoon in 1925, given the number of other schoolhouses that had become deathtraps for the students they housed throughout the day. Though the windows and panes were gone, there was minimal damage to the brick structure and not a child inside had lost his life. The children, however, *were* understandably anxious to get out of that safe building—and home to their families. If there were homes and families left to go to.

The teacher was trembling wildly, Lena noted, as he cried almost helplessly to the departing pupils "'Whatever you do, stay in the road!!'"

*   *   *

Adrian Dillon felt the same sense of uneasiness.

"At first that teacher wouldn't let none of the kids go home," he said.

Once the boys from the now-canceled marble tournament had gotten inside, they were stuck inside, it appeared. Willard Carlyle was determined to try and keep complete control of the chaotic situation, even though he continued to speak in contradiction of himself.

"I've never seen anybody so nervous," Adrian noted. "He didn't want us to go, but he kept saying something about staying in the road if we were *going* to go."

*   *   *

"He kept telling us that," Lena related. "See, there were a lot of old wells back then behind everybody's houses, and he didn't know but that they were laying there bare, and we would be picking our way along and not paying attention and fall into one.

"But what road?!" she offered in a rhetorical question. "We couldn't see the roads. There was so much trash and rubble laying all around, so many houses flat, that once we got outside the school building, we couldn't get our bearings."

Weeping schoolchildren then took to the 'streets,' looking for some indication that their lives, which had been spared by the sturdiness of their school, would go on in some semblance of normalcy. Striking out across the landscape that had been laid to waste in a matter of minutes, it seemed as though normalcy was something that they may never know again.

*   *   *

John Dillon rose to his feet between his two friends who had met their death at the hands of the crazed killer winds that had swept across the plains with swiftness unlike anything anyone in Parrish had ever beheld.

"He had a piece of two-by-four embedded in his thigh," said John Dillon's son Wendell about his Dad. "I was always fascinated by that part of the story. When I was younger, I spent years with the understanding—and this was my imagination, you know—that Dad had this huge chunk of two-by-four, and it was sticking out of his leg all big when he got up from out of that rubble. Really, it was about six inches long and about an inch-and-a-half wide, and he could walk with it, and later he went and had it surgically removed.

"He kept it for awhile," Wendell mused, "but I don't know what happened to it.

"Anyway," Wendell continued, "he got himself up and started walking the mile back north to the farm house."

On the way, another Dillon, a neighbor and distant relative by the given name of Carroll, met John. Marv Wilson related what information was transferred from one man to the other:

"Carroll was coming back from north of town, where the storm had hit really hard," Marv said, "and Grandpa said to him, 'Did you see anyone back up there (at the home farm)?' and he said to Grandpa, 'Aw, Johnny, you don't want to go up there, there's nothing left.'"

"Carroll told Dad, 'They're all dead!'" Wendell summed up the conversation, amazed at the other Dillon's presumptiveness and audacity in such a horrible situation.

"Some neighbor," Wendell added.

*   *   *

"I didn't see Dad when we went home," Adrian Dillon said. He, his older brother Leonard, 13, and his younger sister, Ruie, almost 10, had left the Parrish School together when teacher Willard Carlyle finally let them go.

"We high-tailed it home," Adrian said, "and we came up to a high-power line that was down, and knew we couldn't get past that by ourselves. So a neighbor man there, he helped us through this, and we finally got back to the house."

There was literally nothing left of the Dillon homestead. Adrian remembered the response of the young Dillon children, who had survived the assault of the storm by being in the schoolhouse:

"We came to the house and it was gone. So we thought, 'Oh, everybody's dead.'"

However, the remaining Dillons, Faye, Wendell and their mother Edna, were indeed *not* dead but coming to, relatively unharmed, in a field adjacent to their farm. Their house was blown away; all that existed, insisted Marv Wilson's grandmother Edna Dillon in later years, were the front door of the house and a table, which sat in the field with them.

And here, Adrian noted one point that might have escaped the memory of one 12-year-old, except for the fact that it was a point of life or death, and something the Dillons had observed as they 'high-tailed' it back through town to their home.

"That building me and the boys ran into when the storm had come up, that Crisp Store, was blown down, and several were killed," Adrian said.

The photograph's original caption, etched by the photographer, reads ""The way Wm. Biggs saved his life at Randall Smith's farm near Parrish, Mar. 18." William Biggs likely wasn't the only one who employed this method of hanging on for dear life in the Franklin County village – and points before and after – that was decimated by the Tri-State Tornado.

It was a good thing for all, he indicated, that they had abandoned their marble games and had at least gone to the school when they had.

∗   ∗   ∗

All around the residents of Parrish, their homes were demolished and many were burning.

The miners at the mines closest to the storm's path later reported they had felt the shuddering of equipment underground, and knew something dreadful had happened. They came up out of the mines, Lena Ross' father included. Her father traversed the road from Logan mines seven miles to the northeast, down to Parrish, in a mad run, intent on locating his family.

Many returned only to find the village in ruins and, for several, the nightmare they had feared as they ran all the way, amidst trash and debris—the battered bodies of their loved ones, some blown far from their homes or businesses. The killer storm had finished its grisly work.

"There was a field east across from the Baptist Church," remarked Marv Wilson, "called Cooksey Field. Grandma said there were quite a few bodies blown up to that field (as it was north of the tornado's track), but that was where they started taking the bodies from town so they could be laid out for identification."

Lena Ross had a different take on the field.

"I saw it," she said, her voice small and quiet. The fine lines etched around her eyes deepened; she didn't want to talk about it; that was clear. But she was compelled.

"It looked like there were 300 bodies laying out there if there was a one. Everybody was bringing the dead people to that field. And even after they thought they'd got 'em all, the farmers kept finding them, later on, when they were getting the ground ready for planting. And that always made me think of that field."

In a child's imagination, 'many' bodies could conceivably look like '300'; it was unfair to compare numbers when the dead were being laid out for claim by family members. In all, Parrish suffered the loss of 22 residents; many bodies were picked up and carried to Cooksey Field by the winds. Most were carted there by men grieving from the seats of wagons that carried the carnage and the women sobbing uncontrollably over broken bodies of their children and other loved ones.

It was estimated by some that a few bodies may very well have been blown there out of West Frankfort, that some of the 148 dead from that city ended up in Cooksey Field as well. Regardless of how they got there, they left an indelible image on the minds of many of Parrish's survivors, which included a total injured, but not fatally, of 60. These survivors now beheld their town in ruins, as the beast continued its run across flat prairie land, avoiding established villages and communities, but creating more destruction on a record scale in its wake.

And the city of Parrish was dealt a death blow; never, claimed its residents in the years following, did it recover from the killer storm that visited their village on March 18, 1925.

The little community stood in mute evidence of this contention. The blood-red, white-trimmed shotgun building in the center of town, formerly a filling station and grocery store, was, in late 1999, one of only four structures that remained of many which once served a thriving community in the earlier part of the century.

"In 1925, you had prosperity like you can't imagine. Coal miners, other people working…" Adrian Dillon offered. "In the same place I'd play marbles, I'd see miners shooting dice for $100 bills.

"Then, in 1929, the Great Depression came along and took away what the tornado didn't," he said. "One of those same guys I saw shooting craps came up to my house one day and asked if he could borrow my hound dog and take him hunting, so he could get a rabbit to eat, things were so bad.

"After Logan Mine went down, around 1927 or '28, and the Depression, well...things were just bad for Parrish."

For Adrian Dillon, the tornado took more than just his home. It took something else that, to a 12-year-old boy, meant an awful lot.

"I was a great marble player," Adrian asserted. "I had flour sacks of marbles when that tornado came, had 'em all in my bedroom. I never seen a one after that. I was a good player. I played for keeps."

And so, judging by what little was left behind, did the tornado of March 18, 1925.

Gone were the other two groceries. Gone were the clothing store, the post office, and the train depot. Even the railroad tracks were gone. Left behind were residents who knew the identity of the killer of '25 as something their grandparents talked about, and a little lady named Lena Ross whose house had the railroad running behind it.

Chapter Twelve

Rural Area—Parrish to Crossville
Hamilton/White counties, Illinois

The next stop on our road trip had already been made—and that one had been home.

It was home that sweltering July weekend because we were back in our old stomping grounds, and those grounds were just one county north of the disaster area: home was Wayne County in Southern Illinois. The storm had been to the immediate south of us, in Hamilton and White counties, decades ago. It made itself busy with the task of taking the lives of so many from those two counties.

It was hauntingly only the elderly who remembered it with accuracy. Those younger than their late sixties were constantly confusing that 1925 storm with so many others that had swept through the vicinity in their lifetime, having only heard of the storm in passing. What it was like to experience it, for the most part, was kept to those who had suffered quietly with its memory.

I was determined to discover those of my neighbors to the south who had gone through it, and to tell their stories and bring the attention at last to them, and to the obscure little places that the Tornado of 1925 screamed through on that March day, creating devastation on a scale not seen in the many decades since.

After the storm tore Parrish apart, leaving only fires and floors of the little businesses and bodies of the residents lying in Cooksey Field, it spun out of Franklin County and bolted across the fields and forests into Hamilton County. As reports from that time attest, the storm roared into the vicinity of a tiny community called Braden and skirted a few miles southeast of a little village called Hoodville. I found those two communities located in a vast expanse of land across the center of Hamilton County that was amazingly flat and marked with the occasional criss-cross of little creeks and culverts, which often flooded out after torrential rains.

At that moment in its history, the tornado was at one of the widest points in its relatively long life. Accounts from the area, and the results of surveys performed by experts in the weeks and months after the storm, noted the damage path was more than a mile wide across the flatlands, measuring in places as a mile and a half.

One could only imagine the horror of the black cloud, choked with debris from the cities and villages it had sucked up and decimated, as it blasted across the plains of Hamilton County. Weather-wise farmers were no doubt in awe of the approaching menace, yet not so awed they were frozen and incapable of action. If that action were only to gather their families and head for a safer room in the house or to the storm cellar which only a few possessed, then that was what they did. They could not control the forces of nature and they knew that; but faced with the enormity of the storm as it shrieked toward them, they wondered if they controlled anything at all, even if it were only the actions required to preserve their own lives.

Contrary to popular belief in the decades following the storm, the "Rural Area—Parrish to Crossville" of Hamilton and White Counties was not a lightly populated, uninhabited frontierland. Farms dotted the land to the right and left of every accessible road in the county; in fact, the density of population on a comparative scale to the end of the twentieth century was considerably higher in the 1920s as to that of the populations at the turn of the twenty-first century. Farmers were the mainstay of the economy where mining had lost its industrial grip beyond Franklin County; however, a few stray mines here and there could be found in the days of 1925 in both Hamilton and White counties. As mentioned before, farmers grew what they consumed and sold what they did not consume, making enough monetarily and preserving enough materially to get by from year to year. As well, families tended to be bigger in the days of 1925 as opposed to later years; there was no average two-and-three-quarter kids per household, but more likely five or more children in the farm family. All this promoted the fact, as well as the appearance, that the strike area of the killer tornado during the forty minutes or so before 4:00 p.m. was a bit more heavily populated than most understand it was.

At a forward moving speed of 60 miles per hour, with winds later estimated to be still exceeding the 250-290 mile per hour mark within and in front of the vortex, the horrific storm slammed across the countryside. Only by the grace of God did the raging weather narrowly miss established villages and towns as they existed within the two counties. Passing south of Hoodville also meant the storm passed well south of McLeansboro, a much heavier populated city and the county seat of Hamilton. It was the only location in

Hamilton County that wasn't considered a village or a primary farming community, and it would have been another Murphysboro or DeSoto or West Frankfort were the tornado to have spun its way through a few miles north of its actual path. To the rural residents of Hamilton County affected by this intense storm, however, it was no consolation that the twister did not strike the population center of McLeansboro.

The damage it created was indeed enough in the rural area alone. For several of those farm families in the area in 1925, the human tragedy hit too close to home. Many lost not only farms and livelihood, but parents, siblings, children and friends. Some survivors had scars that were caused by flying or falling debris. Most carried scars that surfaced only when the skies grew dark in the southwest and the rumble of thunder could be heard and felt just before the wind began to blow.

> *"They picked us all up, other people on the way to Enfield, too, live and dead in the same wagon, but sorted us all out…"*—Clarence Wilson, *age 5 on March 18, 1925*

One of these who carried scars was Clarence Wilson. He lived six and a half miles southwest of Enfield (which put him in Hamilton County, just across the county line) at the time of the great tornado. In his later years a resident of Carmi, Clarence, who was 80 in 1999, had no difficulty recalling the events of that March day, even though he was not quite six years old at the time of the great storm.

Like many of the survivors, Clarence spoke of the disaster as if it had happened only yesterday. But he was quick to point out that his perspective had been questioned in the past by others, most of whom, it turned out, weren't there to witness what he had witnessed and really had no right to question at all. Still, he said in defense of the survivors and their families to whom they had related the tale, "Everyone may have their own point of view (on the storm)," and, while this was true, it did not diminish the impact his story had on those who heard it.

"We were in Hamilton County," Clarence began, "in the country in a 'thicket,' a whole bunch of timber in a tree line next to the house. So there was no way of seeing the storm as it came up."

The Aaron G. Cloud House built in 1884 on the west side of the courthouse square in McLeansboro became the city library in 1922, three years before the tornado. Although the Tri-State Tornado did not hit the county seat of Hamilton County the building was used as a temporary holding place to determine how badly injured the survivors were, and which of them needed to be transferred to area hospitals. The building is now the McCoy Memorial Library.

The defensiveness Clarence carried was apparent in his story as he continued.

"People have said 'Why didn't you get away from it?' *That* was why; because it moved so fast, and because we couldn't see it 'cause of all the timber.

"Plus, why, there was no cars out there, there wasn't but a couple of horses where we lived. But even if we'd had a car, we couldn't have got away. The storm just came up too fast."

Clarence recalled how the sky in the distance "looked so bad," but not enough to accurately portray what was hidden within the fast-moving clouds.

"My dad was the one who noticed something wrong. I was looking out the window when Dad was getting us all into the house. Then we could hear it roaring, and when I was looking out the window, well, we had an old surrey (wagon) that was out in the drive and it just went up in the air, and I could really hear the wind roaring." The storm was upon them then, and "the glass in the windows blew all inside the house."

By this time Clarence was away from the windows, after having seen the surrey go straight up with the force of the winds; he knew that whatever was about to happen was very powerful and that it was wise to move back into the center of the room.

"All my family was in the same room, and I got under the bed. My mom and dad and three sisters, two older and one younger, were in the room when the walls fell in. My baby sister was killed. She was eighteen months old, and a wall fell on her.

"There had been a man who had come to visit his daughter in the area, and he was staying with us. He had been there about two weeks. Well, they found him in the woods later; a tree fell across his neck and it killed him," Clarence said.

He had no recollection of who the man was, where he was when the winds took him, or of the one who finally found the body. Clarence knew only that the man had been staying with his family, and suddenly, he was gone, never to return.

The Wilson family suffered greatly in the disaster, both physically and emotionally.

"My mom's arm, shoulder and collarbones and hand were broke up— scratched. My dad had a leg and hip broke. He was on crutches forever after that.

"And me, I had a skull fracture, and I was all boogered up, broken bones in my back and neck. See, I was knocked unconscious for three or four days, so I don't know but from what people tell me."

What Clarence did know was that neighbors from the south, and out of the path of the storm, came to see about his family right away.

"The neighbors came through," Clarence said. "Quick as they cleared the way, they came through with a wagon and started to haul us into town. They picked us all up, other people on the way to Enfield, too, alive and dead in the same wagon, but sorted us out, you know."

Understanding that the extent of his injuries kept Clarence unconscious for so long could sum up the trauma of being so severely injured. However, the trauma of losing a family member needed only to be heard in Clarence's words about the dead and the knowledge that both living and dead were transported in the same wagon.

"My baby sister still works on my mind some," Clarence admitted in a hushed voice.

Once in Enfield, the bodies were taken to the Sam Orr Mortuary.

As for the injured, "They had two doctors there in Enfield, and they took care of us.

"But there was nowhere for us to go except to an uncle in Enfield. And we stayed there until the end of summer. We rented a house later on that fall. It was hard," Clarence said.

While the Red Cross was noted to have provided a considerable amount of aid to storm victims "they didn't help us one penny," complained Clarence. "Those higher up in society got help from them, but we didn't. The neighbors just helped."

Amidst all the tragedy, there were a few humorous notes Clarence was pleased to share.

"Later that winter we heard about some livestock that was most likely ours that managed to show up in places where they couldn't've got unless the wind had blowed them there. In Trumbull, west of Carmi, there was a guy with a vacant lot and up came three extra horses to his lot, so he let 'em in with his horses, but he wondered where the three extra came from." Clarence's father later heard of the extras in the herd and went to check them out; sure enough, the horses, fully intact and still useful as transportation and work animals, were theirs, and he lead them back home.

"And we heard that several of our cows went to Dahlgren. I could say that they walked, but they weren't there after the wind came, so we'd figured at that time that they were dead. Come to find out they were just in another place; wind must've put 'em there.

"I'd had a goat who had two babies who *didn't* get blown away. They all three came bleating around wondering where the house was after the storm."

Clarence said that there was no trace of the house after the storm moved on.

"You couldn't tell where the house had been," he insisted. "But we also had a lake there at the house; and after the storm, the entire lake was gone— wind just sucked it out. And we'd had a cistern at the house and a well at the barn, and there was no water in either one of them afterwards for a long time."

Clarence carried a six-inch scar on his head from a skull fracture, and while, after many years, the fear had diminished considerably, he recalled how it was for his family and others around him who had experienced the storm in the years following.

"Us kids were scared to death when it came up a cloud," he admitted. "We used to cry and take on, cling to Mom and Dad. But we grow up. It's passed over the years.

"I had to tell myself for awhile that if it was going to happen again, it would, and there wouldn't be anything I could do about it. But even though I'm not scared any more, I still remember how it felt."

*   *   *

After our initial July 1999 road trip, we took many more over the rest of the summer, fall and winter, the Hamilton/White counties area included. During these we discovered several graves, some in obscure, unexpected locations, all marked with the date "March 18, 1925."

Two sites in particular stood out to us: one, Vernon O. Miller, was in the cemetery just south of Enfield. It was yet another small, gray, polished granite stone with the date of Vernon's birth, Dec. 13, 1912, and the date of his death, March 18, 1925, side by side. The young man was only 13 years old. In several months of searching, I did not come across anyone related to the boy, or anyone who knew him. It was as if the only remembrance of him, other than in the mind of God, would always and only be the stone, and that was a very lonely thought.

The second site was in a very isolated cemetery that sat across the road from the Albion Cumberland Presbyterian Church just north of Norris City. The church, while indicated by a sign off Illinois 45 South, was difficult to find and was actually recommended to me by gravediggers whom I met at the Enfield Cemetery after finding Vernon Miller's grave. They were certain, they'd told me, that there were at least a couple of graves with markers bearing the date of the tornado in that particular cemetery.

As difficult as it was to find, it was worth the drive, for the land and forests in that part of White County were incredibly lush and beautiful.

The two stone markers, soft marble and faded with the passing of time, stood facing the east, as was the tradition in Christian-based religious graveyards. They belonged to William S. Hollister, born August 9, 1887, and Rossetta E. Hicks-Hollister, Born May 31, 1875; each having perished in the storm of the mid-1920s. The silence of the graveyard and of the stones reflected the information that I was able to obtain about the Hollisters.

Nothing.

It would likely have taken plat maps and serious investigative work in order to discover where the Hollisters had lived, what they were like, who their children were, and what they all did for a living. I could find no one in the Norris City/Enfield/Carmi area who seemed to know anything about him or her. The gravestones bearing the date of the tornado were elegantly carved with a gate, supposedly opening into heaven, and the inscription, 'Come Ye Blessed.' It was more disquieting than comforting; to make matters even more so, someone had, perhaps several months before, placed a very small nosegay of flowers at the base of each stone. The flowers had faded in the summer sun and elements; so had, it seemed, almost everything about the Hollisters except their marble monuments.

*   *   *

As the storm crossed into White County, it traveled north of Norris City and south of Enfield, leaving behind demolished farms and death. A somewhat hilly area of western White County did not slow nor affect the storm as it approached the city of Carmi, devastating what lay north of the city limits.

The area that became Carmi had seen its share of natural disasters, enough to do for several lifetimes. The killer storm of 1925 was about to add to that list.

But in December of 1811 and again in January of 1812, massive earthquakes struck from out of the New Madrid Fault Zone (and possibly the more recently-discovered active fault line, the Wabash Valley Seismic Zone). The sleeping giant that is the New Madrid is a system which runs southward from Southern Illinois through Missouri and into Arkansas, beneath five state lines, and cuts across the Mississippi River in three places and the Ohio River in two. The fault line that shifted in the early 1800s ran along what became the southern part of Interstate 55 through Missouri, from Memphis to St. Louis, and up. On those dates in 1811-1812, it reached its hand from the bowels of the earth, rattling the countryside like a dog viciously shaking a rabbit.

The earthquakes, which would have ranged in magnitude of between 7.8 and 8.2 on the Richter scale were it to have been in use at that time, turned soil into liquid surfaces, temporarily shifted the course of the mighty Mississippi, and swallowed Native American settlements whole.

According to historical accounts by white settlers there, one of these was reported to have been beside the Little Wabash River near Carmi. Legend had it that the sand upon which the encampment sat experienced liquefaction and the village disappeared. To this day, a ridge runs along the prairie between Carmi and Maunie, which is a little community on the banks of the Wabash. The ridge was either driven up during the quake, or formed when a plate beneath the prairie sank in the tremendous force, leaving the ridge as a sort of step down onto the prairie from the north to the south. Few knew about the nature of the ridge, even those who drove daily over the blacktop that crossed it. The fact was, no one was alive who could recall the events of those days in the early 1800s.

However, many were alive to give accurate accounts of the mighty storm that swept death across the countryside in March of 1925. Several still lived in and around Carmi; some lived outside the city at the time; all recalled it as a time when they thought the world, as they knew it, was going to be swept away from around them and that nothing would ever be the same again.

Far too many were correct.

"I felt the barn raise up three times and on the third time, it fell down flat"—Berniece Dartt, age 8 on March 18, 1925

The daughter of Ed and Carrie Winter, Berniece Winter Dartt was one of seven children in 1925, a member of a large family who had been in the Carmi-White County area all their lives. The farm upon which they resided was not far from the Hadden School building, which was a one-room school that stood two miles north of Carmi out Centerville Road.

As were so many of the survivors, Berniece was a mere child in the third grade in March of 1925. Her recollection of that day, and of the days that followed, however, was incredibly keen and precise, for it was a day she has remembered all her life and would remember until the end of it.

"My dad and grandpa were coming home from a sale that day," she said, "and stopped by in a two-horse buggy to pick me up because Dad knew I was done with my lessons for the day and that I didn't have any overshoes to walk home."

It was about 3:45 in the afternoon. The day was looking more blustery and stormy, and Ed Winter felt that rain was imminent. The two-mile walk that his daughter was used to taking would be a difficult one in the rain that threatened, so he made the effort to avert a potential problem.

"My brother, Edwin, who was ten, was to stay at the school," Berniece explained. "His lessons weren't done yet, and I guess Dad figured that he would get home okay, or else he would come back around and pick Edwin up before it stormed really bad.

"The rest of my brothers were staying home that day because they needed to help Dad on the farm, and Edwin, even being as young as he was, was going to be expected to do the same."

So the Winters took the buggy and headed off down the road to the south in order to reach their home before the weather would overtake them.

> *"So the doctor looked at me… and pretty much threw me in the corner and said, 'We'll bury him in the morning'"*—Ralph Gates, age 4 on March 18, 1925

Just to the south and east of the Winter's residence was a house owned by the Gates family.

They were the consummate farm family, with a two-story frame home and a large brick chimney that ran up the side of it. A virtual orchard of trees surrounded the home, with several outbuildings housing the equipment and animals vital to farm life.

Ralph Gates, lifelong resident of Carmi, remembered the fine two-story house that he and his family—parents, two older brothers and one sister—lived in at the time. And for being not quite five years old in 1925, he remembered considerably more than that.

"My parents had taken me to the Frieberger School for awhile so they could go into town," he claimed, noting with interest the fact that the school folks were kind enough to let a little four-year-old (more than a year younger than the age children typically started school) stay at the school for a period of time so his parents could get errands completed.

"They came to pick me back up and I went in to the house to change into my play clothes," Ralph said.

The area of the house which Ralph regularly changed clothes was situated with the large chimney, and a fireplace outlet built into the room from it, stretching up to the ceiling and on outside the house. Ralph thought nothing of the usual routine of getting out of school clothes and into more sturdy play clothes to end the day—nor of where he was standing when he was performing this childhood ritual.

"But then it got real black (outside)," Ralph recalled, "and my dad was trying to get horses and cattle into the barn."

The nightmare storm continued in its determination to wreak unfathomable havoc on farming communities in these isolated areas—and true to its incredibly vicious persona, the killer weather event targeted the Gates home.

Ralph paused for a moment, and his speech became uneasy.

"And then my mom was sticking pillows into the windows that got broke by the wind. That's the first and last thing I remember."

*   *   *

"Grandma was waiting to open the gate at their house when we got there," Berniece said. "She was waiting because she could see a really black cloud rolling off in the southwest. We'd seen it too. Dad had took up the lines and whipped the horses up fast.

"We could see things in the cloud, in the air," Berniece remembered.

What kinds of 'things'?

"Trees," Berniece said with not a little awe. "Lumber.

"Oh, kid, we could see all kinds of things rolling in the air in front of that cloud. It scared us to death."

Berniece paused for a moment and pulled herself away from the thoughts of what might have been rolling in that cloud, and continued with her story.

"Grandma said to Dad, 'Ed, you stay here,'" Berniece said. "And Dad said to her, 'No, I need to get to Carrie and the kids.'"

Berniece's grandpa got out of the wagon and joined her grandma as they walked up the lane to their own home. Ed Winter and his daughter Berniece quickly drove on.

Carrie Winter and her other children, baby Marie, Herman, age 6, Chester, 13, Carl, 15, and Lena, 18, were waiting in the house and watching in anticipation of the evil dark cloud that could not only be seen but felt and heard wailing its death cry upon the horizon as it closed in on the area north of Carmi.

They evidently took cover, Berniece surmised, because no one was hurt badly, and because they did not see the approaching wagon carrying Ed and his eight-year-old daughter.

"Dad put me in the feedway," Berniece explained, "when we got up to the barn."

A feedway was a narrow walkway between two outbuildings such as a stable and a barn. It was used to travel between the two, and had doorways to deliver straw for bedding, as well as hay and other feed products, from the adjacent building over to the animals within. Apparently Ed Winter thought this narrow feedway, as opposed to being under the larger part of the barn or stable structure, would be safer for Berniece in that the beams had less space to fly or to fall and hurt his precious daughter.

He left her there with a strong admonition: "Don't you dare go to the house!!" he warned Berniece, and he went to undo the horses from the buggy.

"He thought I'd be safe," she said of her father. "He knew I couldn't walk against the wind. He was right. He couldn't even walk against the wind himself. The wind blew him into the woods."

And once again, the report of the three giant gusts of wind surfaced, as Berniece detailed, "I felt the barn raise up three times…and on the third time, it fell down flat."

*   *   *

Ralph Gates' memory stopped at the point of his mother stuffing pillows into the broken windows of their home because the chimney, nearby where he had been changing clothes, came crumbling down upon him, denting his skull and knocking him unconscious.

What he told next was later related to him by his parents, culled from what they could piece together of their experiences.

"My dad was saved by a barn sill that came down on him and kept anything else from crushing him. My eight-year-old brother was out in the yard playing, and they found him under the house with his leg all twisted up. Well, my dad got all of us together, and my brother and sister at the Frieberger school up the road, and took us all into Carmi to Grandpa's, and the doctor came there."

'Triage' was not a phrase commonly used for medical assessment and treatment back in the 20s (in the civilian world, anyway, as it came to be commonly used on the French battlefield during World War I), but the philosophy that controlled it was the same: Victims were divided into three categories—those who are likely to live, regardless of what care they receive; those who are likely to die, regardless of what care they receive; and those for whom immediate care might make a positive difference in outcome. In such a major catastrophe as the great storm, those who were more severely injured, and likely would not survive, were given cursory treatment, limited mostly to comfort and relief from their pain, and the doctors moved on to those whom they felt they could save. The unconscious four-year-old Ralph Gates was one of those unfortunate patients.

"My mom had been almost scalped," Ralph continued, "something caught her in the head and almost tore her scalp off. So the doctor looked at me, pulled out a bottle of iodine and poured it on my head, and pretty much threw me in the corner and said 'We'll bury him in the morning.' Then he went about stitching up my mom."

*   *   *

"What saved me was I was on the inside of the door where Dad put me, and it dropped me right down beside the concrete foundation and timbers fell over me in such a way that I was not crushed," said Berniece Winter Dartt.

This was probably the exact intention her father had for putting his daughter where he did. Short of blowing the two large and well-built structures completely away, there was little that could have harmed Berniece in the narrow, sturdy location where she stood.

Berniece said little about the length or duration of the monstrous cyclone, not even that it passed as quickly as it arrived, but at a forward moving speed of three times the average tornadic storm, the presumption could be made that it was as quick an event as it was catastrophic. The next thing she remembered was the way she was left after the barn collapsed.

"My legs were pinned against horses," she claimed.

She could easily have assumed that the horses, the lifeblood, farm working animals and secondary transportation for the Winter family farm, were all dead—but with her close proximity to them in the collapsed barn, Berniece knew that this was not the case.

"They were knocked out," she remarked. "All ten of them. It didn't surprise me. Everything was flat."

Her father, who had been blown into the woods, came back and instantly began calling for his family, to assess their condition—or whether they were there at all.

"The house wasn't completely demolished," Berniece said. "But the chimney was torn out; the windows and frames were blown out; and the house was blown off the foundation and shifted quite a bit.

"Our Overland car was blown out through the chicken yard."

Carrie Winter and her children were practically unscathed.

"Mom had got 'em all in a room together," Berniece revealed, "and that room held."

Her father, who didn't know whether or not his daughter had listened to his admonition to stay put, didn't want to believe that she was in the wreckage that was now the barn. His first assumption then was to run to the rest of his family, hoping Berniece had made her way there, as they were all safe and intact.

"Dad came running up to them and said to Mom, 'Where's Berniece?!'

"Mom didn't know what he was talking about; she hadn't seen us come home together. She said, 'She's in school!' to Dad; she was in a blind panic."

Winter knew then that his daughter had indeed stayed where she was told, in the barn and that now, in a heap of rubble, there was the most likely place to find her.

"So they went out and called and dug and found me under the pile of timbers with all the horses. I was able to walk to the house after they dug me out. The only thing I had was bruised legs."

Her grandparents, though, did not fare as well at the hands of the storm's violence.

"Some neighbors from about a quarter mile away," said Berniece, "came running and said, 'Ed, help us, your mom's pinned under the house!!' and Dad looked toward Grandma's place and the whole thing was flat."

Ed Winter sent one of his older sons on a neighbor's horse into town to retrieve a doctor, but the boy couldn't get into town due to high water and trees in the road. The land toward Centerville was one of the lowest lying flood plains in that area of Southern Illinois; the rains, heavier in the area of White County than they were almost anywhere else following the monster storm, had made short work of the flood areas causing them to be impassable. Trees were down over much of the road, even though there was a grove of trees adjacent to the homestead that wasn't touched in the horrific winds.

Her grandparents were injured severely in the collapse of their home.

"Grandpa had internal injuries. Grandma's arm had been crushed; they'd had to amputate it a week later. Grandpa died, and while they were burying Grandpa, Grandma died. If they'd had the right medicine, she might've lived. They had no antibiotics."

And for brother Edwin, who had been left at Hadden School on that fateful afternoon?

"The wind blew the teacher, Ernest Lamp, and Edwin into a field about a quarter mile away.

"They weren't really hurt."

*   *   *

Obviously Ralph Gates was not buried in the morning, as the doctor treating the injured that March day had carelessly predicted. Though his recovery was long, his fortune, and that of his family's, was better than some who survived to tell about the day the killer vortex from the sky dropped down upon White County.

"We lived in a tent for a while after that, the Red Cross gave it to us. Then we built us one of those Sears and Roebuck houses because there wasn't anything left of our house but the floor. That new house wasn't a big fine two-story house like we'd had, but we were grateful to have a house.

"We saw a lot of weird things that that storm did. There was one barn up by the schoolhouse that the tornado picked up and turned around and set it back down. Didn't wreck it. And there was a large woods north of the house that had white oak timber in it, and the trees were laying in every direction, no one way they fell, and sheets of tin were drove into them."

Sadly, although the Gates family did not lose all their livestock in the storm as some did, nonetheless the storm claimed some of their animals in the end.

"The animals were banged up and in bad shape but they could've been doctored. And we couldn't doctor them because they couldn't be put in a pen. There wasn't anything to pen 'em up in or with. So we had to shoot 'em.

"I remember after I got to come home, seeing the chickens plucked clean of their feathers walking around naked with their baby chicks that they'd hatched after the storm," Ralph said.

Why were the chickens not blown away? Ralph didn't know, but he was amused by the memory. It was one of the few points of amusement to be found in such a tragedy, but then, there is often more resilience in a child's trauma than in that of an adult's.

While Ralph also bore a white scar across the top of his head as a witness to the events of that day, storms no longer had the effect on him that they might have had, had the outcome been different. Ralph's son, Richard, later moved with his wife Janice to the rebuilt house on the hill north of Carmi. Having remodeled it, it was not recognizable from its former appearance.

Still, the imagination could kick in when standing in the driveway of the Gates' home, gazing toward the southwest and wondering what it must have looked like to see the most hellish of all storms bearing down across that part of White County.

*   *   *

The monster storm was not quite done with White County; there were more families and more farms and communities in the way of its nightmare run.

Barreling across the Little Wabash River and its valley, the tornado swept up on the small community of Crossville, Illinois. Crossville lived up to its name; in those days as in the latter, it was a crossing point for major routes on the eastern side of Illinois. Highways 14 and 1 intersected at the four-way stop, and until more recent years a rail line ran from Centralia to Indiana through Crossville. Several country graveled roads also intersected from directly north-south and east-west with the main roads in Crossville, making it a hub of activity for many years, especially following the Illinois oil boom in the late 70s and early 80s.

Because the mighty storm only brushed the south edge of Crossville, the village did not get a listed distinction of its own on many of the charts that show its path on that day in 1925. Deaths in the village of Crossville from the storm were subjective depending upon to whom one spoke. Rough estimates were only 'a handful;' none in the village proper were killed. However, those to the south lost homes, farms, and more, because, even in a handful, if one is your beloved, it's one too many…

"We all swore the cellar rose up and down, about six inches, each time a big gust of wind went over us…"—Hal Davenport, age 13 on March 18, 1925

George 'Hal' Davenport lived in a single-wide trailer just a couple of doors away from 'downtown' Crossville. He was a jovial man of 87 in 1999, and a lifelong resident of the village. And he gave a chilling, albeit brief, description of what happened in the minutes just before four p.m. on the afternoon of March 18, 1925:

"We lived on the edge of town, just below the Phillipstown crossroads, and the tornado went just a hair north of us.

"When we saw it coming, we went running to a storm cellar we'd had built outside," Hal related.

The storm cellar was the typical kind built with severe storms in mind— only, perhaps, not on such a magnitude as what the twister was about to display. This cellar had been dug out of the ground in the backyard, and had a concrete dome on top to support the roof. The roof was covered by a heavy layer of dirt over which grass had grown throughout the years. Double doors created the entryway, one atop, one below ground: A single door that opened to the outside led to the stairwell, where another door entered into the underground structure itself.

Into this underground structure ran Hal, his grandmother and grandfather, an aunt and a brother.

"The storm came over," he said, "and my grandpa went out into the stairwell to hold the outside door shut because the wind was so strong that it was whipping open the doors.

"My aunt went to hold the inside door shut because Grandpa needed it shut for leverage. All the rest of us were trying to help hold her still so she could keep the inside door shut."

Hal continued, "Grandpa was using himself as a barricade between the two doors. And I swear, and so did my brother, that while that tornado was overhead, we could feel the whole storm cellar raise up and down about six inches, like that tornado was trying to pull us out of the ground."

Asked if the raising and lowering occurred during three intervals, Hal said, "I'm not sure; maybe so. All I know is, the whole cellar raised up and down."

The Davenports were unharmed; not so their property, which sustained substantial damage, even being on the edge of the storm.

"We had a lot of outbuildings," said Hal, "and we had two or three barns completely down, and the chimney tops were all over the yard."

Of the debated deaths in Crossville, one was certain: Fred Bennett, who resided one-half mile north of the Davenports, was right in the path of the killer storm as it went about its grisly work of destruction.

However, there was still some debate, even among Crossville residents, on exactly how Fred was killed. Many report that Fred was in his cellar, and, during the storm, opened the door to see what was going on; when he did so, an object allegedly came along and severed his head at the neck.

Others maintain the fate Fred met was less gruesome and more common among the tornado's victims.

"Fred was down in the cellar sprouting potatoes," claimed Hal, "and he headed up out of the cellar to see what was going on."

Fred Bennett didn't make it.

"He got killed when the house blew in on him. A big timber came down on him—crushed him."

At any rate, those in Crossville who remembered the storm were growing in age, and some memories were fading, especially if one didn't experience them first-hand, as many in Crossville did not due to the storm passing south of the village proper.

However, the fact remained that it did cut through south of Crossville, the last real community in Illinois to be struck by the mighty twister.

It was moving ever forward at 4 p.m. the afternoon of March 18, 1925, on the path N 69 E degrees, never wavering, and never leaving the ground.

Across the two counties of Hamilton and White once Parrish was left behind, 65 people, a shocking number, were killed, all or most of them farmers or members of farm families. A total of 140 people were injured. The dollar amount in damage left by the storm, for this particular area, was, astonishingly enough, the third highest toll overall in property losses: $1,100,000, showing just how expensive farm property could be in an agrarian-based society.

Property consisted of livestock, automobiles, farm equipment and implements, cropland, timber, orchard trees, farm utility buildings, other outbuildings, and farmhouses. A price tag could not be placed on the loss of life the two counties experienced; everything else could be itemized.

The storm, amazingly, was not finished; its devastation not yet complete, its strength not exhausted. In fact, it was building force for one final assault, this time on another state. For in crossing the next wide river on its course, the Wabash, the killer storm, which had decimated entire villages between its start in Missouri and path through Illinois, was about to take on another area of the young countryside. It was about to earn its name by which it has come to be known to this day:

The Tri-State Tornado.

Indiana

Chapter Thirteen

Griffin, Indiana

There is little change in the landscape from Illinois to Indiana, unlike the transition between Missouri and Illinois where the bridge crossed the Mississippi at Chester. The Illinois bluffs continue on for awhile from the west boundary, but, moving eastward, they give way to flatlands. With the minor exception of a few hills outside Griffin, the flatlands of Indiana, as they are in southeastern Illinois, are the rule.

The hills that ringed the west and ran north of Griffin were known by many different names; there was a set of them known as 'Mumford Hills' after the folks who owned the land on them. The ones that stretched to the north were known simply as the 'North Hills' or 'Upper Hills,' and ran a good piece over the country up to where the Wabash River bent against them in its many turns and convolutions.

The Wabash River had in its time been a navigational dream and nightmare by turns. It descended from northern Indiana south of Fort Wayne, and divided Indiana from Illinois just south of Terre Haute. From there it took a decidedly serpentine course down the border between the two states, bending where the White River from Indiana met it at Mt. Carmel, Ill. It grew wrathful and raging in the flood season, and had risen to incredible proportions many times during the twentieth century. And, like the New Madrid beneath the Mississippi on the west side of the state, the Wabash had an earthquake fault line below it known on the more northerly end as the Vincennes, and in the more southerly regions as the previously-mentioned Wabash Valley Seismic Zone. Little temblors could be felt from time to time in the vicinity of the town of Vincennes in Indiana, Lawrenceville on the Illinois side; and in one memorable event on the 102nd anniversary of the great San Francisco Earthquake (April 18, 2008) a sizable 5.2 a few miles outside of Mt. Carmel to the north and west of that city.

Hopewell and Adena Native American tribes were populous throughout the area in prehistory, and many mounds rose up along that stretch of the Wabash, built long before the Oubache tribe, as the French settlers called them, placed themselves down the shores of the Wabash. No one was certain which tribe it was that buried its dead ceremonially in the suspected mound that stood atop 'Skeeter Mountain,' which was a last, great monument of earth and trees before the land dipped into the Wabash on the Illinois side. It was only known that the 'mountain,' a hill on the Illinois side that may very well have been a precursor to the hills that framed Griffin only a few miles away, was preserved and could not be traipsed upon without express written permission from some government agency that controlled those sort of things.

At the turn of the 20th century into the 21st, Griffin itself could be easily overlooked when passing by quickly on Interstate 64. Most folks traveled the highway headed to Evansville, Ind., or points even further east, like Louisville, Kentucky. The exit to Griffin was even overshadowed by markers that indicated a bypass in the other direction to New Harmony, Ind., a highly rated tourist destination for those who loved history of early settlements in the U.S.

Griffin lay quietly to the north of I-64. During the day, it held a profusion of little white and light-pastel-colored houses and trailers, not a mansion among them all, and a post office, gas station and diner all in the same immediate area, sharing one parking lot. Besides Posey County Co-Op, there were no other businesses. There were no schools. There was only one church. The railroad, which was the lifeblood for many small towns at the turn of the twentieth century, was even gone, its rails and ties having been ripped up. Only ghost signs remained to remind folks that a 'rough railroad track' lay ahead, as one diamond-shaped sign proclaimed; and at the diner, which was named, appropriately, 'The Depot,' there was a blinking railroad light to indicate there was once an active rail line running through the little town.

Griffin claimed about 200 residents at the end of the 20th century. That was a little more than half of the amount residing in the town in the early 1920s: 350. At that time, it was a viable town with businesses and trades; grocery stores, blacksmith shops, undertakers…

Griffin had lost none of its beauty over the years, however. At night, streetlights twinkled in perfect rows like gemstones set within a giant, beautiful brooch, each one doing its best to out-glow the other. The town was quiet; and everyone knew everyone. When someone was sick, there was immediate concern; when someone excelled, it was cause for celebration. When someone died, it was a time of mourning, for in each life there was tied up a piece of the village's history that could no longer be repeated and, in some cases, no longer remembered once they were gone.

People moved into the area and moved out again, disillusioned with the quiet and interpreting it to mean 'there's nothing to do around here!' Transients brought a stir and a bit of consternation to Griffin; they breezed in, did their thing, and breezed back out again. It wouldn't much cause a ripple in the scheme of things, but the old timers noticed it, and when the transients moved on, life would get back to normal.

On the afternoon of March 18, 1925, there was a transient that breezed into the area and out again with dizzying and deadly speed, so much so that sixty percent of Griffin's population, almost two-thirds of everyone in the town and on the outskirts, were affected in one of two ways:

They were either dead, or lay injured, some close to death.

It was upon its approach into Indiana that the tornado, now in its third hour of existence, made its way into the record books for speed, duration, and deadly force. Just after crossing the Wabash and just before demolishing Griffin, the horrific storm took a deep, fetid breath and made a mad dash ahead, and increased its forward moving speed by an additional thirteen miles per hour. It barreled into the tiny community at a mind-boggling 73 miles per hour, so fast that its effects would otherwise have been felt momentarily, but so powerful that, judging by the damage to the town of Griffin, some scientists at the end of the 20th century with their advancements in meteorology, tentatively (and oftentimes argumentatively) placed the storm within the imaginary category of Fujita scale F6—'inconceivable winds.' It was possible the vortex winds may have reached well over 318 miles per hour, the cutoff speed for F5 on the original scale (see later chart on enhanced Fujita scale post-2007), and the breaking point into the next category that had not been recorded in weather-keeping history.

Griffin bore too many similarities to Gorham, on the other side of the neighboring state, to let such coincidences go unnoticed. The power gained by the crossing of the river was one; while most weather experts and meteorologists did not contend that crossing a body of water would increase the impact of such a storm, the locals argue the point vociferously. They agreed, however, that the appearance of such a storm, if not the strength, became more menacing. It was by virtue of the fact that water, being driven ahead of the storm by the force of the winds ahead and within the vortex, can create deeper shadows within the debris field and make the tornado appear, in turns, both white and black, wicked and looming in the heavy bruised and battered sky.

When crossing a river, especially one that didn't have a particularly deep riverbed like the Wabash, much more can be drawn up than water. Whatever wildlife was teeming within comes up with the water; fish, frogs, fowl and, shell creatures could be deposited many miles away when the winds were through playing with the aquatic assortment in the atmosphere. The phenomenon of 'raining fish' no doubt came from such a storm; it didn't have to be one as powerful as the beast of March 18, 1925.

Mud and silt can also be drawn up into the vortex and deposited in places that could least use them. The Wabash had always been a rich runoff river. Farming the neighboring acreage always turned up fertile soil that, in rains and subsequent floods, poured into the river bottom and lined it with amazingly fine, black mud. This, too, could be swept up in the grip of a vortex and deposited in some very unwanted locations further along the storm's path.

Another similarity to Gorham's misfortune that Griffin bore was the distinction that of almost 400 in the population, a devastating 60 percent were left dead or injured. Gorham had the second highest death percentage toll, that of 41 percent, behind only Griffin, which took the ignominious first place of percentage of population dead or injured in the tornadic disaster.

And yet a third similarity remained the damage toll. Both Gorham and Griffin suffered destruction or damage to every building and structure in the town and outskirts. Property losses, totaling only $375,000 in 1925 dollars, amounted to 100 percent destruction when the storm's path cut through the little town and all was said and done. Every building bore scars left behind by the violent winds of that March day. It was as if, when crossing the rivers, the tornado drew a huge, angry breath, and in a mighty exhalation, sent it out over the unfortunate town that lay in its path.

To have heard the stories up to this point was heart-wrenching; the mortal terror, the human agony, the blinding pain of young ones losing parents, or of children being torn from the arms of mothers and flung so far and with such velocity that little broken bodies were all that were found, littering the countryside, if any were found at all, in the hours following the storm. The skies darkened once again, but this time with nightfall. The day's moderate temperatures had cooled to seasonable degrees, as the beast bore down on Griffin.

Because of it all, the memory of this particular storm, commonly called the Tri-State Tornado after having reached its third state, stood out in the minds of those in southwestern Indiana, even in the villages hit by it after it moved on, by one name:

The Griffin Tornado.

> *"...I heard what I thought was thunder, or at least thought I heard it. I think I might've felt it more than anything"—Bob Simpson, age 9 on March 18, 1925*

Until early 1999, Bob and Vera Simpson, 83 and 79 respectively, lived in one of the pretty brick houses that sat a nice yard distance off the main street in Griffin. Bob Simpson liked to hang out at The Depot in the afternoons, and he always brought a big Styrofoam cup of coffee home to Vera, his wife of many years. They lived in Griffin all their lives until April of 1999, when they moved to Mt. Vernon, Indiana. They did so even after being lifelong residents of Griffin and unable, as they would express, to imagine any other life for themselves outside their little town.

They independently described a Griffin of 1925 as a frontier town that, twenty-five years after the turn of the century, had become well established in the southwestern Indiana area, including to the southeastern Illinoisans as well. While there was no Interstate highway nor a road for carriages and cars to cross over, Webb's Ferry ran across the Wabash River day and night, even after construction on a bridge began in 1966. As well, the rail line, known as the 'Indiana Hi-Rail,' ran across the river from Griffin to Grayville (Illinois), up to Browns (where it was known as the Illinois Central), and on to Olney and Newton. On the Indiana side, the rail ran down to Mt. Vernon, Ind., and on to Evansville, a burgeoning metropolis in the early 1900s.

Bob and Vera Simpson had a framed painting hanging in their living room of one of the general store buildings of downtown Griffin at the turn of the century. In the painting, many men stood lounging around the front of the building; horses were hitched to the posts lined up off the street. Most of the men wore holstered handguns, as was the practice in the early 1900s. There weren't a lot of wild animals to necessitate carrying guns back then, Simpson pointed out; but there were a lot of wild-tempered men, prior to and during the prohibition era.

Griffin, being a depot on the Hi-Rail line, was a successful and growing place in the spring of 1925. Bob Simpson recalled the 'bread train' rolling into Griffin straight from Mattoon, Illinois. It contained hundreds of three-foot-by-eighteen-inch breadboxes, full of fresh-baked bread, still warm (this, Simpson remembered fondly,

Bob and Vera Simpson

went on in Griffin until about 1940). The 'bread train' likely rolled into Griffin on the morning of March 18, and people were going about their business several hours later at about four in the afternoon, slicing their fresh bread, receiving children home from school, preparing the evening meal, wondering what crops were going to do well that year and what to plant in the garden.

If the day was ready to grow wrathful and deadly, none of the residents of Griffin knew it. They continued on with their daily routine as though nothing of any monumental import were about to happen—because as far as they knew, nothing was.

> *"...he said that the wind was blowing him straight out. He was almost blown away"—Ellen Vanway Nottingham, age 11 on March 18, 1925*

On the southwest edge of Griffin, across from what later became Interstate 64, was a cozy little farm owned by the Vanway family. Kell and Martha Vanway had been making a good living in Griffin, and together had raised six children.

Ellen Vanway

Harry, 15, was not their eldest child, but was the oldest at home that day. He was followed by Ellen, 11, Hellen, 9, and Evelyn, 8. Not home on March 18, 1925, was son Lilburn, 18; Lilburn had gone into Evansville, Ind., in search of a job. His older brother, Aubrey, 21, was living in Michigan at the time.

The Vanways had seen their share of heartache in the time they had lived in Griffin. Many years before, prior to the birth of the two youngest girls, Martha Vanway had given birth to a baby girl, also named Helen, who had died at six months of age of diarrhea. Martha Vanway had loved the name Helen, and, with a slight modification of spelling, had named her next daughter Hellen, in memory of the beloved baby they had lost.

On March 18, 1925, Harry, Ellen, Hellen, and Evelyn Vanway were riding the school bus home. Theirs was the first stop on the circuitous route the bus ran initially, south and west of the town, before it went back to the school and picked up a second load for a north and east run.

There were about twenty children on the bus, noted Ellen Vanway Nottingham, who later in her life became a resident of Evansville. The bus was not quite to capacity, but it seemed full with energetic, rowdy kids on their way home to start the second leg of their busy day.

"Harry and another boy, I don't remember his name," began Ellen, "always rode in the back of the bus. There was a little rail and some stairs that went out the back of the bus, and they would ride there and look out the back. When we were riding through town, we saw Mom…she'd gone to the store there, Fisher's, and had also gone to the post office."

Martha Vanway watched the school bus pass by and waved from outside the post office as she stood next to the family car.

"Of course, she didn't think about it," Ellen remarked sadly, "but that was going to be the last time she was going to see Harry. She always said he looked so happy, hanging on to that rail in the back of the bus, waving to her."

*   *   *

In the Hyatt household, which sat just outside the village, little Vera Hyatt, four years old, had just seen her father, Sidney Hyatt, off at the door, on his way out to track down the town doctor.

"My brother, Lucien, had the measles," Vera said of her younger brother, who was home in bed, considerably ill with the virus. "Dad was going into town to talk to the doctor about what to do with him. And we were all waiting on my older brother Dale to get off the school bus."

The Griffin school bus was a common sight to villagers in the mid-twenties. Due to the growth of the community, and the fact that, as a predominately farm community, families and their children were spread out all over the countryside, one large, two-story school housed almost all the school kids in that section of the country—and they were bussed in from the bottomlands and farm prairies over that part of southwestern Posey County. There was the school bus to accommodate them; according to reports, not a nice modern school bus, but an old-timey, bus liner type of vehicle, capacity about 30, which carried children outside the village area where it was deemed a little too far for them to have to walk. It was on its first route into the country, the one going south and west, the route that headed straight into disaster.

Dale Hyatt, reported his sister, was one of the fortunate ones.

*   *   *

"I was living with my grandma in the Upper Hills," said Bob Simpson, "and I went to a country school up there, the Wabash Township School, so I wasn't in Griffin."

Simpson, who was nine at the time, was not unaware that there was something in the air. Something ominous was approaching. The day had been a little misty and damp in southwest Indiana; a late afternoon shower would not be unexpected by those in the area.

But a late afternoon shower wasn't exactly what the day had in store for Griffin.

"I was almost home," reported Simpson, "and I heard what I thought was thunder, or at least thought I heard it. I think I might've felt it more than anything."

Sitting in the school wagon, he looked over his shoulder and saw a gathering darkness in the southwest sky over the river. It was amazingly black,

Devastion in Griffin, Indiana, where 26 people died from the deadly storm.

Simpson said. And it was moving with amazing speed.

Vera Hyatt, her mother Lilly Jane and little brother Julius watched with astonishment as the twister began to destroy the outer edges of town from the southwest. It looked to be headed to the northeast and would likely pass into the hills, where young Bob Simpson had made it off the 'school wagon' that delivered children to and from Wabash Township School. He had made it to the home of his grandmother, Mary Elizabeth Simpson.

Somewhere on its south and west route of the loop, however, the Griffin school bus, emptying itself of children, was in the general vicinity of the south end of the Upper Hills. But the bus was not alone. Killer winds were looking for innocent victims in those hills. As it trundled along, the windows of the bus were opened outward, letting in the warmer-than-usual air.

*　　　*　　　*

The Griffin school bus rounded a corner and was facing the southwest as Chick Oller, driver, opened the door to discharge his first riders—the Vanway kids.

"There had been things flying around in the air before we got home," said Ellen. "Trash. Bits and pieces of things, I remember. We didn't know what it was."

The confusion was no doubt compounded by the relatively clear sky in the moments before the gathering darkness began sweeping across the Wabash. For minutes, the last ones of clarity many would have, it appeared as though there was going to be a simple rainstorm—until the ominous sign of fluttering paper, cloth, branches and other debris began pelting the stopped school bus.

Reality struck—there was nothing simple about it and *It* wasn't just another rainstorm.

"Mr. Oller was just sitting there, and we were getting off the bus," said Ellen, "and were all going inside the house when we looked back and saw the cloud coming out of the southwest."

Chick Oller was frozen in shock, witnesses later reported. He was so taken by surprise that he could only gape at the cloud. His mouth hung open as the storm barreled across the river and fertile farmland, brutalizing the earth as it plowed into Griffin.

"It looked like big black smoke rolling on the ground," Ellen said shakily. "It was rolling on our ground in the pasture to the southwest. I didn't see a funnel; just rolling clouds like you see in the sky only it was on the ground, like black smoke."

Ellen's older brother made one summation as he reached the back of the house, turned, and saw the monster approaching from the southwest.

"Harry said, 'Oh, that's some cloud!'" his sister related.

> *"Oh, my, everything was just wiped off the town. It was pitiful"—Ed Westheiderman, age 14 on March 18, 1925*

Ed Westheiderman, unlike Bob and Vera Simpson, was not a lifelong resident of Griffin upon his interview in 1999. His father, George Frederick Westheiderman, had been a blacksmith in New Holland, Illinois after the turn of the 20th century. George Westheiderman moved his little family to bustling Griffin when he heard from his brother, also a blacksmith in the town of Griffin, that he had more work than he could handle in his blacksmith shop. So they moved to Griffin in 1914, when little Ed was but three years old.

His modest home was on the other side of the block from where Bob and Vera lived; it faced toward the river and offered a splendid view of the sun setting over the Wabash and the south part of the hills enclosing Griffin on

210

two sides.

Ed, a little sickly at times in his later age, made a brave attempt at recalling the day of the great tornado that swept his hometown. He began his story in a rather bemused way, as if he was fully aware that there was nothing anyone could do about the outcome of the weather on that day, nor of the death and destruction it brought with it.

He was firm and struggled little with the beginning of the tale, about how he was already home from school by 4 p.m. and was playing with his brother, Paul, eight, and sister Mable, seventeen, and a neighbor's daughter, whose given name he did not recall. He was able to remember that she was ten or eleven years old, and that her surname was Young.

"It was simply a rainy-like day," Ed said.

He indicated they were all being just normal, active kids, burning off excess energy on a Wednesday afternoon.

"We would run into the house when it was raining enough to get us wet. Then when it would quit we'd go out—and Mom would yell at us to come back in."

Mom, Pearl Westheiderman, looked up from her late afternoon chores and noticed the sky in the west growing terribly dark, almost like nightfall. The black wall of clouds, reported Ed, hurried across the sky and horizon, blotting out all the light from the afternoon sunset. They could not see it from the east side of the house where they were playing—but they could sense it creeping up behind them. For all they knew, night had descended upon Griffin two hours early.

It wasn't night that was descending, however. It was something much more ominous.

"We happened to be inside when Mom saw how terrible it was," Ed said. "I don't know how she knew, but she did. And it's a good thing she did."

Pearl Westheiderman frantically called the children inside and close to her in the kitchen.

> *"...about a quarter of a mile away, I saw corncribs... and they seemed to explode"—Kenneth Woods, age 11 on March 18, 1925*

Ed Westheiderman had a roommate, Kenneth Woods, a former principal of the high school which was in nearby New Harmony for twenty years. He could lay claim to being a lifelong resident of Griffin as well ("Except for when I was in the Army," he noted).

His father, William Woods, was a grocer in Griffin during the early twenties. He owned and operated not only Woods' Grocery but also a café within the building. The store was on the main drag of Griffin, and it was a busy place. Railroad workers, area business people, and farmers alike all congregated there and smoked, drank coffee, and debated the times as they were transpiring around them.

"School had just been dismissed," said Kenneth, who was nine at the time of the Griffin tornado. "I always walked up to the store from school. That day, my teacher, Fern Delashmint, walked with me.

"We had just walked into the store when it got black all of a sudden…very, very fast."

One of the unusual elements of the storm, as has been noted, was its incredible speed. Imagine something the size and strength of the monster tempest approaching, but imagine it approaching at the rate of a modern eight-cylinder automobile speeding down an Interstate, eight miles over the posted speed limit of 65 m.p.h. At 73 miles per hour forward moving speed, it is almost incomprehensible to imagine such vicious tornadic activity. But what was even more incredible was the fact that the wind speeds, for whatever reason, were keeping equal footing with the forward pace of the storm. For every increase in forward movement, the wind velocities increased within and surrounding the vortex. Upon its approach to Griffin, weather experts estimate the Tri-State Tornado was bellowing devastating wind speeds well in excess of 300 miles per hour. Damaging enough on their own, the winds became an unholy terror when combined with the speed of descent the tornado took as it crossed the Wabash River.

The damage this kind of hellish storm caused was, and remains, unfathomable.

"I happened to look out the window, and about a quarter of a mile away, I saw corncribs…and they seemed to explode. The tops would go up," Kenneth reported as some unease crept into his voice and as he used his hands to indicate the movement of the cribs exploding, "and the sides would collapse."

The debris was blown, he indicated, to splinters. Some of it appeared to evaporate completely in the twisting vortex of the monster storm.

But oddly, "I saw no funnels," he remarked.

While most Griffin residents, some long gone, had always reported a single, large funnel, there was at least one set of bone-chilling stories about the presence of possibly *more* than one funnel that might contain a modicum of truth to it, and maybe beyond—as multiple tracks, like those found in Biehle, Missouri, were located by disaster officials after the tornado's passing.

Legend had it that, there in the Upper Hills where the view was best, many observed not one, not two, but *three* twister tails descending the west hills that framed Griffin as the storm raged ahead from the southwest. One report actually indicated that as the mighty storm crested the hills, the twisters dropped one after the other, One-Two-Three, and began a dance of death as they swarmed around the village of Griffin, repeatedly combining, twisting apart, and regaining their hold on one another again, as they consumed the little town. The report may coincide with the strange anomalies within other reports, beginning in Annapolis, and continuing on through DeSoto, West Frankfort, Parrish, Carmi, and Crossville, about the 'three-gust' effect witnesses felt or experienced. In Griffin it was confirmed, by disaster reporting sources in the weeks following the storm, that three funnels were observed between Griffin and Owensville, Indiana, for a measured, professionally-surveyed track of six miles. That the tracks were present was indisputable. Obtaining eyewitness accounts of the three twisters, however, was a more difficult task. It was one of those things that must just be accepted, since the ground survey bore it out; scientifically, it has yet to be proven as to just exactly what, and how, the multiple vortices did what they did to the ground. Scientists know that there is a difference between the "improbable" and the "impossible."

To Kenneth Woods, the impossible was happening in front of his eyes.

"We heard an awful loud noise, a roaring sound, and the whole front of the store just blew away.

"My father and another customer at the store, Mr. Williams, had been standing at the front of the store," Kenneth said. "But when it passed, they weren't standing there. I had to go look to find them."

"The tornado of March 18, 1925? I wasn't born then. But it almost kept me from getting here!"—
Charles Price, not yet born on March 18, 1925

Charles Price, who lived in Evansville, Indiana in late 1999, wasn't necessarily a survivor, but sure knew enough about the Griffin Tornado of 1925 to tell his tale about it. He was one of the few people interviewed who listed off the date of March 18, 1925 immediately, unlike many other survivors, most elderly, who sometimes, while not getting the year wrong, would forget the date. He knew, he claimed, his history.

Even though "I wasn't born then," he offered, laughing.

"But that storm almost kept me from getting here!"

Explaining the odd comment, Charles went on. "My parents have told me about the Griffin Tornado time and again. My dad, William Price, was uptown when the storm struck, and my mom, Mae, was home with my older brother, James, who was about four years old, and my other brother little Billy Jr., who was only three or four months old. I hadn't even been thought of yet; I was born a little over a year later."

Both parents, he reported, were standing outside in the manner of the curious and unaware, looking at the approaching darkness from the southwest; his father from a location uptown, his mother from their home. Of all the things he remembered his parents telling him, the main thing was about the speed of the massive storm. Therefore it didn't take long for normal curiosity to rapidly turn to terror.

"They each always said it was so incredibly fast," Charles recalled. "Dad was uptown, and he ran home, it was only two or three blocks away from where he was."

"But by the time he got there, the house was gone."

Mae Price had run back inside their home to the two small children when the violent winds swallowed their house. Previous reports indicated that Mrs. Price ran to the crib to pick up baby Billy, Jr., but the powerful storm struck at that moment and swept both of them away. However, that was not entirely accurate.

Baby Billy was not immediately swept away by the winds, nor was his mother. Mae Price evidently reached the crib and the baby, and snatched him up in her arms as the storm's winds began screaming and clawing at and around them.

"Mother always thought she squeezed him so hard that it killed him," said Charles, "but I went to check on death certificates (in Mt. Vernon, Ind.) several years ago, and it listed 'trauma by blunt force from flying objects.'"

The remorse Mae Price was unable to disguise each time she spoke of the death of her infant son was not lost on Charles Price William as he grew up in the shadow of what happened that day. Charles pointed out that his mother could rarely bring herself to remember that aspect of the nightmare. But remember she did, and related the rest of those terrible moments to Charles at least a few times in his life, enough for him to be able to give an accurate description of those moments when the house came down around the Prices, bricks and timbers beating and battering them all.

"My older brother James was injured badly, on his legs and ankles," said Charles.

He noted that because of the fact that it was years later he found out about the tragedy, he could only surmise what the extent of James' injuries were.

"I'd just heard he was under some bricks," said Charles. "He wasn't in school at the time, he was just a little guy, three or four years old.

"Mom had five fractures in her skull. In one of them there was a big piece of wood buried in the side of her head and she had a concave area there from then on."

*   *   *

"It didn't occur to us to get into the fruit cellar under the kitchen," said Ellen Vanway Nottingham. "We'd never experienced anything like that before."

In fact, Harry Vanway, instead of taking cover, began to panic when he saw the cloud rolling along the ground to the south of their home. He took off—back out the front door, straight toward the storm.

"He was running across the front yard when it hit," said his sister. "Something struck him in the back of the head and he went down."

Their father, Kell Vanway, was busy in the barnyard when the school bus pulled up to the house. When he turned to see the storm crossing the river bottoms behind him, his first instinct was to run. But he had seen that his children were already in the house, Ellen said. His thoughts were with them and the fact that he felt they were safe inside. Instead of going to them, his daughter Ellen said, he responded with self-preservation.

"Since he knew it was going to storm bad, he laid down and held onto a fence post in the chicken yard.

"Later he said that the wind was blowing him straight out. He was almost blown away."

Ellen Vanway herself was in a state of utter confusion. She had seen only her brother go running out the front door; she didn't know where her sisters were, or if they had even made it in from getting off the school bus. So she went into the dining room, wandering almost, and while she was there, she "could hear all the windows crashing in.

"I was knocked down and could feel things hitting me," Ellen said, "and I couldn't get up."

The school bus, with driver Chick Oller helpless at the wheel, was hit with the force of a freight train. The bus tumbled back, over and over on itself, and was spun upright for several moments. The storm began tearing at its sheet metal body. By the time the tornado was through, the bus lay shredded, upright in a nearby field.

"The only thing that was left of it," said Ellen, "was the flat bed and the wheels. And two neighbor girls, Helen Harris, who was twelve, and Ethel Carl, about the same age, were under the back two wheels, one girl under one wheel on each side."

Dale Hyatt, Vera Hyatt Simpson's older brother, managed to escape the wrecked bus with a broken arm and only a few minor scratches. He sat in stunned silence watching the killer storm speed away, and that was when, by reports, the agonized screams from the injured began.

*   *   *

When the storm passed over the Westheiderman household, Ed, his mother Pearl, and siblings Paul and Mable, along with a neighbor girl, were "wadded up together" in the kitchen.

"We got hold of each other," Ed said, "and we were okay, we hung on. It blew most of the house around us away. It was tore up. But there were two rooms left, enough left to protect us later.

"I don't know where we would've landed if we hadn't been there in that spot. Nobody was hurt."

In the bizarre manner mighty tornadic winds often have, the force of the blow managed to miss the inhabitants of the house and moved on—but it did not miss the blacksmith's shop in town, where Ed's father George Westheiderman was at work.

"Dad was in the shop and thought he had enough stuff around him to protect him when the wind went over," said Ed.

But it was the surroundings of the blacksmith shop, and the materials and tools of the trade, that created the predicament the elder Westheiderman found himself in by the time the winds had passed.

"He had a lot of equipment and parts inside the shop, and what happened was this—the wind drove a steel pipe between the two bones in his left arm and spread 'em apart," said Ed.

The piping was evidently small enough to fit into the space between the radius and ulna bones of his forearm. But that was not the worst of the injury.

"Then it fastened him when it was driven into a tree that had fallen over into the shop."

That fastening was just the beginning of the horror. It was bad enough to be pinned like that—but the pipe was long; according to Ed Westheiderman, while a good bit of it, probably several inches, was driven through the arm and into the tree, the entire length of the pipe was ten to twelve feet.

It didn't faze George Westheiderman, who was panicked but determined to get loose and home to his family.

"He was trying to get out and see how we were. So he worked that pipe out of there," said Ed.

Ed wasn't talking about out of the tree. He was talking about his father walking along, away from the tree, and *pulling his arm off* from the impalement of the steel pipe.

"He worked it off, and he got free," said Ed. His father was a blacksmith, and as a result, insisted Ed, "he was strong to begin with.

"His arm was wrapped up for quite some time but he did alright," Ed said. The point being, with the elder Westheiderman, what his son's next comment was: "And he never gave in for anything."

*   *   *

Nine-year-old Kenneth Woods, having ducked under a counter in his father's store when the killer storm swooped down from the darkened heavens, emerged and went to look for his father as the winds were subsiding.

"I found him," Kenneth said, his voice sorrowful from the recollection. "The front of the store was gone. Dad was lying under bricks and wood. His chest was crushed. That other customer, Mr. Williams, was up there with Dad. He'd had his head cut off from the front. It was lying backwards, hanging by threads."

Such horror facing the child no doubt shocked his senses to numbness—

but not enough, said Kenneth. He noticed details contained within the damage, such as that of another creature who had survived the disaster.

"We had a collie dog who stayed with us in the grocery part of the store," Kenneth recalled. "He was under the counter somewhere with me, but where we stored sacks of flour. The one thing I remember after the storm was that it rained terrifically. The dog had been under there with the flour, and when he got wet, he was just plastered with that wet flour."

The doughy dog aside, Kenneth went out of the structure where the storefront had been located ("We never did find the east part of the store," Kenneth noted). "I was so stunned, I just started walking away," Kenneth said.

Teacher Fern Delashmint was unhurt and still in the vicinity of the store when she saw Kenneth wander by.

"My teacher said to go back to my relatives," Kenneth said. "But when I got there, our house was destroyed."

*   *   *

Charles Price's father, William, had run all the way to his home in Griffin to find, not the house destroyed, but completely gone. His infant son, William, Jr., was also gone, the winds finally sweeping his battered body away, they would discover later, into a nearby field. Four-year-old James lay nearby, injured in his legs and ankles. Not far away from them was their mother, with severe head injuries and punctures.

"Dad went to Mom's parents, the Stallings (her father Charles, was a town undertaker) hoping to find them alive there so they could help, but they were dead too. So when he went back toward where our house used to be, he found Mom between the two houses, and he went to get some help."

But those who came to Mae Price's rescue, including her husband, discovered what they thought to be a corpse instead of a living woman—and they loaded her up with the rest of the dead, in a wagon, and shipped her off to a building in New Harmony (the Ribeyre School Gymnasium), which was providing its facility as a temporary morgue.

Due to the extent of her injuries, it was unlikely Mae Price was aware that she was being transported with corpses destined for a final resting place. Her youngest son had nothing to offer about any remarks Mae may have made regarding the horror.

But he did know "They pitched her in with the dead there in the gym. And then someone noticed that she moved. So they went to work on her."

Still, Mrs. Price did not escape the infamy of having been mistaken for a corpse in the macabre shuffle of bodies from Griffin to New Harmony. At least one popular account of the storm from 1978 has Mrs. Price dying in an open location next to her infant son; the *Evansville Courier* newspaper, in its first reports of the dead, listed her as deceased.

William Price Sr. was devastated, reported his youngest son, Charles. The elder Price's experience, however, grievous though it was, was symbolic of the perverse and unpredictable nature of the fierce storm.

"He came home," said Charles Price, "his house was gone; his son, dead; his wife listed as dead; her parents were dead. And all he got from the storm was a little scratch on his back."

*   *   *

"I really thought that the world had come to an end, and I was lost," said Ellen Vanway Nottingham.

When asked what she meant by 'lost,' Ellen explained: According to what she had been taught about religion, when the world ended, there would be a judgment for the good and the bad, and all around them would be destroyed. Being a little girl of only eleven, Ellen wasn't sure into which category she fit. She knew she was alive, and that she had then gotten up and was looking around. But nothing was as she remembered it. The house was gone. A few remnants of the farm buildings stood, but barely. Worst of all, there were others within her line of sight, and they weren't behaving normally, either.

"I could see people standing all close together, where the bus had been," Ellen said. "Their faces were black. I didn't know any of them, and they all started running across the field toward home, I guess. I don't remember anybody saying anything—and then they took off."

There were no words being said, but there were tormented screams, and crying, however faint. From the corner of her eye, Ellen saw her father approach a still form lying on the ground in front of the house. It was her brother Harry.

Ellen didn't know what hit her brother, or what exactly it did to him, only that it struck him in the back of the head, and he was killed instantly, lying facedown in the just-greening grass of the front lawn.

Her father told her to get back, and, no doubt when he ascertained there was nothing he could do for Harry, he arose quickly and went to tend to the living.

"His head was all nicked up," Ellen said of her father, "and he had a finger torn off his right hand, the middle finger, I think. But he gathered two of us girls and got us together, so he could go get help for us."

The other two girls, Ellen said, were nowhere to be seen, but somehow, her father located at least Evelyn and put the two of them, Evelyn and Ellen, in one place.

"There were two big trees blown down in the road," Ellen said, "and he put us next to those maple trees, found a piece of a wet quilt, put it over us and told us to wait while he got help."

The girls had no intention of doing anything *but* wait.

"We didn't say a word," Ellen said. "We were in shock."

"Evelyn's leg was cut to the bone on her knee. I didn't see it, but she was on crutches later for a long time, and the scar was really bad."

Kell Vanway took off toward what was left of the town. Ellen made no comment about what he might have observed of the utter and complete destruction on his trek across the three miles back into Griffin, nor that he told his children anything at all about it.

Martha Vanway had been trapped inside a store in downtown Griffin.

"She was in Fisher's store," Ellen said. "They'd all gone to the back of the store when they'd seen the storm approaching. Bill Fisher, the owner, was trying to hold the door open or shut, I don't know, whichever way the wind was blowing, and Mother had her hands on his back trying to hold him so he wouldn't blow away. But the store owner's nephew, who worked there, was killed."

Martha Vanway ended up being trapped by fallen bricks and debris, inside the store.

"She couldn't get her foot out without losing her shoe," said her daughter, "so she got that shoe off and she went walking, she couldn't use the car, you know, downtown was just destroyed, but she walked those three miles with no shoes on, in just stocking feet."

Martha Vanway was a mother, after all. She knew she'd have to get back to her children, no matter what the obstacles or difficulties.

"She said there wasn't a house standing between town and home," said Ellen.

"When Mom got home, she found Hellen, and she took Hellen's clothes off. She had to cut them off."

Martha Vanway got down to her daughter's bare skin. "When Mom got to Hellen's skin, why, there was a big gash in her side. She threw up her hands and said, 'Oh, my!'

"There was debris, straw in that gash…" Ellen's voice trailed. "There were some people who lived over from us and they had a farm; their buildings weren't too bad. So we started there but we couldn't get any farther than the bridge. So they took us to a wagon and that's where Mother was when she found out how bad Hellen was hurt.

The little girl, the family's second Helen, died before 2 a.m. that morning.

"The doctors said all the doctoring in the world couldn't've saved her," Ellen said, but it was evident in her voice that the post-death observation was of little consolation to the family.

*   *   *

"My brother Alan, he was 19, he was just barely hurt. But Mom, her back was broke," said Kenneth Woods about what he discovered when he made his way home from his father's grocery store. "Now, we didn't know her back was broke at first. People came to help us and loaded some of us on the Hi-Rail train and took us to Olney. They took Mom in, but I didn't know what they were doing to her right then, because they thought my jaw was broken and they put a wire contraption around my head and made me keep it on 'til morning. But my jaw wasn't broken. It was bleeding, but it wasn't broken."

As it turned out, Bess Woods, Kenneth's mother, had a broken back, but the first hospital they took her to didn't seem to think so. Out of deference to the hospital's reputation, Kenneth chose to not reveal which one in the area it was.

"But she stayed there better than a month before they figured out something was wrong. Then my sister Margie took her to Indianapolis where she lived, and to St. Vincent's Hospital there, and within ten minutes Mom was diagnosed with a broken back. There were two or three broken vertebrae back there, and she was in a body cast for almost two years because of the misdiagnosis.

"The thing I remember the most is that it was dark when they loaded us on trains that night, it took so long for the trains to reach us."

Sadly, Kenneth Woods' aunt was Mae Price's mother, Alice Stallings, whom William Price Sr. found dead, along with her husband Charles, in their collapsed home.

Part 1 of a panoramic photograph of the destruction in Griffin.

"I remember that they found Aunt Alice," said Kenneth. "She had a two by four run through her stomach.

"Uncle Charles was also killed. He was the town undertaker."

*   *   *

"Our house was just north of the railroad tracks," said Ed Westheiderman. "It was the second house north of the main part of town on the east side. Harry Young, the depot agent, lived next to us. That one girl, his daughter, was with us and she was okay in our kitchen.

"He had another daughter who was killed."

Ed's voice got tight at that point, the memory of the horror dawning afresh as he said, "We didn't know about storms then. We were so young, we didn't know what was going to happen."

Ed's confusion as a child of 14 was evident in his voice as he continued:

"People just…didn't know what to do…people came to help and you didn't know who to trust," he said haltingly, considering each phrase in almost a bewilderment of how to rightly put it. "You gathered clothes right after and kept 'em in a bundle."

Yet, the magnitude of the storm and the damage it inflicted was beyond incredible to youth such as Ed, and the awe in his voice, even at the horror, was unmistakable.

Part 2 of a panoramic photograph of the destruction in Griffin.

"Buildings downtown, and we only lived one house from the business area, were burning that night. To me, it was something exciting, the fire. I mean, we didn't know," he explained.

What Ed didn't know was that like in Murphysboro before them, people were burning alive in the fires that raged through Griffin following the storm. The flames were high and huge and leapt into the night sky. Reports from Crossville claim the heavens were orange in the northeast by the time darkness uncomfortably settled in for the night.

Within the consuming flames was the body of Vera Hyatt's father, Sidney. He was crushed to death beneath a chimney and his charred remains were eventually found in the days after.

"Sometimes you think about it and sometimes you don't like to," said Ed, with tears running down from his eyes. He apologized gently. "I'm sorry," he said. "I don't do this very often. But…

"Oh, my," he sighed, "everything was just wiped off the town. It was pitiful. We were running everywhere, not knowing where we were going."

Ed began to address some of the issues that were of primary concern to the people of Griffin in the months and years following the storm.

"That school bus—we found out later about it—it wasn't the driver's fault." He addressed the situation of surprise many found themselves in as the

Part 3 of a panoramic photograph of the destruction in Griffin.

rumbling, death-dealing mass was approaching. "People were just standing out and looking at that thing coming and looking their last." He sighed once again, and choked back tears with the memory, sobbing. "You've pulled this out of me. I remember so much.

"But none of us in my family were hurt, we got in that corner of the kitchen, so we weren't hurt."

Then Ed summed up the entire few minutes of the tornado roaring overhead as he felt compelled to add one more injury the wrath of the storm inflicted on its victims:

"Well, only...our feelings were hurt."

*   *   *

There were many stories surrounding the days and weeks and months following the storm's visit to Griffin. There were the tales of impalements and dismemberment, something the children of the town for the most part (perhaps excepting Kenneth Woods looking at Mr. Williams' headless body) did not observe firsthand. The children of the village were all that remained in the latter part of the twentieth century. Many stories died when the adult survivors began passing away in the years following the tragedy.

Some stories remained. It was commonly known that many people wandered about Griffin in a state of shock, unable to get their bearings or find their relatives, who were unrecognizable to those very relatives looking for them. In the winds, their clothing was ripped from them, and river mud, that fine, silky silt lifted into the air by the storm as it crossed the Wabash, was driven into even the smallest pores in their skin so they looked no longer Caucasian but Negroid. There were tales of people picking up any scraps of clothing they could find to place on their naked bodies, little girls dressed in boys' clothing and boys dressed as little girls. Parents actually overlooked their own children in their wandering and utter chaos and confusion, mistaking them for the ones that lay dead in makeshift morgues. There were tales of the hungry wandering the night of March 18, 1925, scraping for bits of food they could find in the rubble of what once was their hometown. Pretty little Griffin was left more than thunderstruck. It was literally obliterated.

Regardless of what shreds of humanity it left behind—and in Griffin, it wasn't much—the Tri-State Tornado was not through with its macabre twisting dance of death. It rode the sky out of Griffin, leaving behind a once-thriving town that in the days following the storm filmmakers would descend upon with their giant, primitive reels and cameras to capture the hideous footage forever. The film shows, in jerky black and white, many dazed, still horrified, residents wandering around what used to be streets, looking for what used to be anything that was theirs, anything recognizable.

And, records indicate the massive storm upon leaving Griffin broke into three snaking tornadoes that sped out toward their next destination: Owensville. Somewhere in between, the three took a slight jog to the north, and left their eerily straight N 69° E degree heading to turn nine degrees to the left, once again combining to become one ferocious twister. Had the storm not turned, it may have skirted south of the next town. As it was, a total of 85 farms would be completely destroyed within the next few minutes, as it made Owensville the next entry in the record books and left 60 percent of Griffin dead or injured. One hundred percent of the village of Griffin was damaged or destroyed. Not a building stood that did not experience some form of damage—if they stood at all.

The time was 4:07 p.m. The day of the tornado was reaching its final, destructive minutes, but it was not finished yet. Two more communities lay in its path, and the appetite the beast had for lives and livelihood was by no means sated.

Chapter Fourteen

Owensville, Indiana

Owensville, Indiana is an unusual, pretty and neat little town situated quite in the middle of nowhere. While Interstate 64 isn't far from it, there is no major Interstate or highway system that runs through or around it; however, it lies in the intersection of two little Indiana state highways, 65 running north-south, and 168 running east-west.

From the east entrance into the little town the land was so very flat and almost delicate looking. There were no rolling hills nor heavily forested areas to give it any kind of rugged appearance; fields in every state of farming, from harvesting to tilling to lush green winter wheat made the land seem somewhat fragile on the late autumn day I was there visiting in 1999. The sun was setting ahead of me as I drove into town alone minus children on this trip; the city limits were surprisingly abrupt, appearing from empty fields and beginning with rows of houses and neat yards and fences.

The charm of the diminutive burg rang true of a German village. On the square rose a little park; pretty trees, bare of leaves, still served to make the place warm and inviting. What added to this impression was a sizable library that hugged the grass in the center of the park square. Its stone walls and warm lights shining brightly through the tall windows gave it the look of a college campus building. Through those windows could be seen shelves upon shelves of books inside, lining the walls with their information just waiting to be learned.

As I had found to be the usual case, the sunny, unseasonable day only lent to the feeling of unreality that a storm the magnitude of the Tri-State Tornado could actually have slammed through this homey little town. Recovery over the years was inevitable, the face of the town had to have changed dozens of times over the decades, but in the warmth of the sunlight on that December day in 1999, at about the same time in the afternoon as the tornado struck—4:10 p.m.—it was difficult to conceive of the rolling, pitch black clouds, the hail, the lightning, the terror that visited Owensville on that nightmarish March almost 75 years prior.

A little police department sat on the square in Owensville, along with a flower shop, a café, and some other businesses, facing the library as if in reverence of it. Neat houses, some small cottage dwellings, some ponderous and mansion-like, lined the streets around the periphery of the square, indicating that Owensville was a village of prosperity as well as of humble households. A school sat a few blocks away from the square in one direction, a gas station and convenience store in another, and one of the town's water towers in yet another.

And right under that water tower, seeming cold in the warm winter temperatures, was the cemetery, its monuments rising gracefully and with finality, indicating the deaths of those who were lost to the March twister of so long ago.

Many had heard the legends of what was here still called the Griffin Tornado; some older folks at the café knew of the story but were rather inaccurate in their recollection of date, details, and in particular the number of those injured or killed outright. What devastation took place in Owensville, (to my greatest fear and adding pressure to the feeling of urgency in completing my quest), was growing ever more dim in the minds of those who endured it. And those who had not were now subjected only to the legends, the old stories that were not much better than rumors—that the nation's most incredible tornadic storm had passed over this little town and had left six dead or dying, 47 moderately or severely injured, and a monetary loss of over half a million 1925 dollars.

I began my search at the home of a lifelong resident of the area. And as I continued, I discovered that in Owensville, of all the places I'd visited, the stories were more intertwined than anywhere else; the feeling of community ran high in Owensville, and the residents did not travel far from where they were raised.

Even after the most devastating event to befall their small town almost tore them all apart.

> *"We knew we couldn't get the dog out, we didn't try; it was just pinned up in there, howling. It was awful"—James Richard Armstrong, age 9 on March 18, 1925*

James Richard Armstrong got around pretty good for a man of 83 in late 1999, although he would say otherwise were someone to ask. His hearing wasn't what it used to be, and for this he apologized, but he made up for it by listening attentively and giving thoughtful and revealing answers to questions about the colossal storm. Thinking about it didn't seem to disturb him, but he was also in the presence of his family, two good-looking sons and their equally attractive young wives, and James Richard found himself able to relax as he related the nightmare of that day so many years before, what he felt and what he saw, knowing that in the here and now, he was safe in the stillness of a December day almost 75 years past, family all around.

As an eight-year-old in 1925, James Richard Armstrong (referred to by everyone who knew him, even in his 80s, as James Richard) lived four miles northeast of the home in which he currently resided. The Armstrong family had a pleasant, working farm life; on their property were a barn and granary, and horses and mules ran through the barn lot and adjacent fields, waiting for their workload each day, resting in the barn at night. The Armstrongs tended gardens and grew crops; all in all, they were a typical rural Hoosier family.

James Richard was one of five children belonging to Harvey and Florence Armstrong. On that particular day of March 18, 1925, the elder Armstrongs were in attendance at Bargain Day in Princeton, a citywide sale that had been advertised heavily and was designed to bring out the spring feeling in folks who needed to brush winter aside and begin looking for new items they needed at prices they could afford. The Armstrongs had taken off in a horse-drawn wagon while the children were in school, intent on getting back with their purchases before too long. All day long, James Richard reported, his parents had expressed a feeling that it was going to come a rain sometime in the afternoon, and they wanted to get back before it happened.

"It was okay earlier in the day," James Richard said about the weather, a fact to which so many eyewitnesses attested. He noted that during those mild spring days, any farmer was going to feel suspicious about the onset of afternoon storms when the morning dawns warm and muggy. Those farmers were right; as James Richard continued, "It got to be a dark, rainy day right before the storm."

Music had always played a prominent and pleasant role in Alice Bruce's life. In her home were keyboards, both organ and piano, and sheet music, hymnals and songbook after songbook. There were even wall hangings reflecting the importance of music in day-to-day life.

> *"We just made it to the restaurant. Just as we did, the trees were laying down just like they were tired"*—Alice Bruce, age 16 on March 18, 1925

"I'm a retired schoolteacher," Alice said, "but I've always been in music. I've played piano for church for seventy years and I'm still playing—I picked up the organ in later years."

She was well dressed and well spoken, lively at 90 in 1999 and vibrant in red and black and white, seated comfortably in her home. The phone rang occasionally; Alice listened carefully for the distinctive rings between her phone and a business phone in the kitchen area.

"Right now I'm working with the funeral director here in town," she explained. "I have to listen which phone it is so that I know when to pick his up or when to let my machine get mine."

Alice was a tender 16 years old on March 18, 1925, and was enjoying the full delight of music at that age. In her high school, from which she graduated in 1927, the choir was working on a spring operetta and she was leaving school at around 4:10 that afternoon, only to turn around and attend evening practice.

"We lived north of Owensville back then," Alice Susan revealed, "and Mercile Kimmell, a neighbor of ours, and I were walking from the old schoolhouse that afternoon to go uptown. I was going to be staying in town with Grandma, because that particular evening we were going to have operetta practice."

Grandmother Susan Emerson was boarding her granddaughter overnight; the Bruces were also in Princeton at Bargain Day, like the parents of James Richard Armstrong.

The adults of Owensville were likely more attentive to the weather than the pupils, who were pleasantly distracted by the prospect of performing in the operetta; as Alice said, "The principal had said there was a storm coming. He said, 'Don't delay; go right straight home!'"

However, Alice and her friend had other ideas about after-school, pre-practice activities.

"Mercile and I were headed uptown and were intending to go to the restaurant there," she said. "We had just made it to the restaurant.

"Just as we did, we looked out the windows. The trees were laying down," Alice Susan indicated with a flattened palm suspended horizontally, "just like they were tired."

> *"Before (Mother) made it across the kitchen, the windows were blowing out, and the plaster was falling"—Alice Spindler, age 5 on March 18, 1925*

Alice Field Spindler, a resident of Fort Branch, Indiana in 1999 was an incredibly sunny and pleasant woman, given to kind words and a gentle manner. In her life there may have been plenty of things to cause her to be this way—the love of two parents, closeness of a family, a wonderful spouse later on, good health. But the darkening skies of March 18, 1925, made their impression on her all the same. While the massive storm did not in later years dampen Alice's rosy outlook, it clouded her eyes nonetheless, even though she was successful at keeping the worried tone out of her voice when speaking about it, unlike so many others who endured the nightmare at such a young age.

Alice concurred that the day began bright and beautiful.

"My father, Erastus Field, they called him 'Rass,' was putting a new fence around the garden," she said, indicating that even in mid-March, the weather could prompt the good countryfolk to take advantage of the opportunity to get outside and prepare for the spring planting season right around the corner.

"But it began to cloud up around 4 p.m.," Alice remembered, "when he was finishing the fence."

She recalled the clouds getting darker and darker, building on the horizon and effectively blocking the approaching sunset.

"Now, there were forty acres of woods behind our house," Alice noted as a caution, "and eight acres catty-cornered, just west of us across the road. We didn't see the clouds well because of it—the woods."

The Field family, Alice said, could only see the gathering darkness above the line of the trees—but that was enough.

"Great big raindrops began to fall," Alice said. "Then they changed to hail and rain together. The hailstones weren't real big, oh, maybe like big peas, but there was a lot of them for a little while.

"We had been smoking meat in the smokehouse in the garage that Father had just finished building. It was hogs that we had butchered, and they'd been salted out and were hanging in the smokehouse to be smoked. My mother, Fannie Field, went to the kitchen to get water to drown the fire in the smokehouse," Alice said, indicating that her mother was aware of the impending danger to their house from the approaching storm.

The garage, storage area and smokehouse were connected to the existing structure of the Field home; Rass had recently built on to it. Fannie Field ran with a bucket of water toward the attached structure, but got only as far as the kitchen window, according to her daughter Alice.

"Mother looked out the window and saw the dark cloud. We couldn't really see it good until we had got inside the house, because it hadn't got over the trees yet 'til we were inside," Alice recalled.

But by the time the storm had breached the tree line, it was too close, and almost too late.

"Mother realized what was happening. She started back into the living room to tell Father the cloud looked just like the one his Aunt Ella told them about that had hit Mattoon, Illinois, in May, 1917.

"Before she made it across the kitchen, the windows were blowing out, and the plaster was falling."

*   *   *

Unlike Fannie Field, James Richard Armstrong, at eight, didn't know what a tornado was.

"We were in the school wagon," James Richard said.

Like many to the west in the Griffin area, horse-drawn wagons carrying kids to and from school were the norm, and according to accounts, the Owensville area had at least two such wagons full to capacity with kids.

"We started seeing small limbs fly off trees…then the wind swept us away."

The wagon had just driven past a house on the right side of the road that had a front porch. James Richard remembered being aware of this, and of one other thing: his lunch pail

"The last I saw the dinner bucket," he said, "it was blowing down the road. When the wagon went over, our only thought was to scramble across the road and get on the porch of that house we'd just passed."

Some of the children did, James Richard said; others remained in the ditch nearby where the wagon had overturned, the team driving it still attached and waiting, standing upright but just barely, in the ferocious wind.

*   *   *

"Normally I'd have been in the school wagon driven by John Gee," Alice Bruce said of her regular school day activities.

The wagon driven by Gee, however, was not the one that contained James Richard Armstrong and other children from his area of Owensville; Gee's wagon was a bit more fortunate—it remained upright on its drive home.

"John Gee's wagon escaped by minutes," said Alice. "Just minutes after it passed the Christian Church on Main Street, the church went right into the street. Fortunately the wagon had gentle horses; they weren't easily scared, otherwise I don't know what would have happened to those children."

From her vantage point inside the restaurant uptown, sixteen-year-old Alice Bruce watched incredible sights as they unfolded before her eyes.

"There were some ladies in the basement of that church, quilting," she said. "Mary Fleener and Mary Armstrong Knowles, James Richard's cousin. The church blew away from atop them. They weren't hurt at all."

A very large branch flew up against the door to the restaurant, said Alice, and for awhile the patrons felt as though they were going to be trapped inside the building while the town came down around them. The men in the restaurant went to the door, Alice said, and pushed against the branch, and when it got far enough back out into the street, the wind slapped at it again and took it on its way northwest.

"Then," she said, "they had to stand there and hold the door *shut*."

*   *   *

The five-year-old child that was Alice Field was grabbed up by both her parents and taken quickly to the front door, where the three stood until the mighty storm had blown over.

"I was between them," she said. "We didn't see a funnel cloud, just the dark of it. It only lasted a few minutes."

At 73 miles per hour forward moving speed, it undoubtedly did. The monster storm passed overhead at three times the normal speed of any rotating storm, but the winds were deadly while they were present.

"Our neighbor, Frank Sanders, had been visiting us. He ran out the door and grabbed a large tree limb that had just blown down. See, this was a really big tree limb, most of the part of a tree, and it had heavy branches. Frank was out in the open; he saw the chance to hang onto something that might hold him down until the wind was over. He made it through. He wasn't blown away," Alice Field Spindler said.

"It was a miracle. None of us were hurt."

*   *   *

"The trees had fallen across the road," James Richard Armstrong said. "There were so many down, and telephone lines, too. There was no way we were going to be able to get the team that had been pulling the wagon to get us out of there."

So, the group of children that had been in the school wagon, which was now lying upside down in a ditch, gathered together with the wagon driver and set out walking. There wasn't one among them hurt seriously; the wind had spared them its wrath and had focused only on the inanimate objects as the recipients of its damage.

At least, as far as the group had thus beheld.

"We came off the porch of that house and out of the ditch, and we got going," he said. "Then when we started walking out, and the driver walking out with us, why, just up the road, there was four people killed and kids of that family were walking with us. We went down as far as the barn lot."

James Richard hesitated, barely perceptibly, and continued on with his story.

"There was a dog crippled, howling, having been blown up into the tines of the mowing machine at the King farm. A boy that was walking with us lived on that farm, what was left of it. We knew we couldn't get the dog out, we didn't try; it was just pinned up in there and pierced through with it, howling. It was awful. The boy was devastated. It was his dog."

Things grew worse for the boy, although at first he did not realize it.

"We were going past the King farm. Four of the Kings were killed on their homestead. But their kids walked past and didn't know their parents had been killed," James Richard said. Sadly, "They were the first ones of the group to ask if their parents were okay."

Of course, the wagon driver, the only adult in the bunch, didn't know how to answer that question and just kept walking, four miles, on the way into town.

"When we were walking, we went under trees, over trees, around trees…we were just a bunch of scared kids…we began to wonder where Mom and Pop were…"

The images of that long walk on James Richard Armstrong had their impact years later.

"Everything we saw on that four-mile walk was just wrecked!!" James Richard Armstrong declared.

*   *   *

"My parents were in Princeton, and they started walking home," said Alice Bruce. "They met up with the school wagon that had made it through the storm, and a couple of nephews that had been in there joined up with them."

Alice's single-mindedness in the face of the disaster was somewhat amusing and at the very least, touching.

"I stayed on in town expecting to go to operetta practice, thinking it was going to be okay."

North Main, she said, had been damaged pretty badly. Some houses were destroyed; barns were blown over into roads out in the country. All these things her parents saw and reported back to her; Alice stayed at her grandmother's house, although all the windows there, she remembered, "came into the house. She was a block east of the church, so she was on the south edge of the storm.

"Our home wasn't even touched."

*   *   *

"Our house had a lot of damage done to it," said Alice Field Spindler. "The house was moved back two feet both north and west. All the plaster in the house was blown off and out. The windows were also broken out.

"The furniture was all covered with mud and manure. Most of the furniture was broken."

Alice recalled one odd thing about the damage, or lack thereof, which was sustained in the brunt of the storm.

"My doll was in the box on the library table. A piece of heavy glass was lying beside it and for some reason mother picked it up and put it over the doll. The doll was untouched.

"Mother's dishes were in the three-cornered cupboard and were thrown out into the floor. Luckily most of the dishes were unbroken."

*   *   *

Sadly, such luck was not with the King family on March 18, 1925.

"There were the Kings, and they had a son, who was married, and they had a baby. The son, Walter King, and his wife Lori, were killed outright. Their baby blew back into the woods," James Richard said simply, "and they didn't find him for three or four days.

"Someone went back to our old barn lot a week later and saw a shoe sticking out of the mud.

"They pulled it, and it was Mr. William King."

The man was buried face down in the mud, with only his foot showing. If someone hadn't found him, noted James Richard, he likely would have stayed that way as the mud dried.

*   *   *

"Father was saying, 'I'm goin' to see if I can find the kids,' and he walked toward town hunting for the school wagon," said Alice Field Spindler. "This was the one driven by John Mounts, that my brothers Emery Clark and Ermal were in, and he found them at the northeast edge of Owensville."

It was the school wagon on which one of the pupils was James Richard Armstrong. Rass Field found the horse team, which had tried to break away from the stranded wagon, and which had ended up straddling a tree. Rass eventually came upon the group of students and, gratefully, his sons, unharmed.

"While my father was there, he was told that my grandmother was still under the debris of their house on the west end of Owensville. Grandpa had escaped by kicking out the kitchen door glass. When they got Grandma out, she was okay except for a broken arm," said Alice in almost a little girl's voice.

The King family saga did not end with James Richard Armstrong's account. Alice had more to tell.

"The children of Bill and Lizzie King, their names were Virgil and Howard and Alice King, didn't stop at their house, since their place was destroyed. Their father, mother and brother Walter and his wife had all been killed. They had just returned home from Princeton, but hadn't made it out of the barn where they kept the car."

Still, some of the children were given relief in the form of a safe and familiar place to stop on their trek to find their own homes and families.

"The children walked towards their homes, stopping at our house on their way," Alice recalled. "There were so many asking, 'Did this hit our house?'

"Mother told them no, it didn't get that far.

"Little did she know that it had gone clear to Princeton."

* * *

The people of Owensville were inarguably entwined by the damage that the storm inflicted, but they were intertwined even before that. The fact that so many of them crossed paths with each other on that day, in the way that they did, is testimony to the fact that community, family, and sincere care and concern for one's neighbor can go a long way to insuring survival against insurmountable disaster.

The Tri-State Tornado, which had been on a perfectly straight N 69° E heading for almost 200 uninterrupted miles, had, outside Owensville, given way to the forces of nature mightier than itself and bent its path a bit. It veered nine degrees north; that fatal move spelled doom to the misfortunate King family, for if the tornado had kept its heading, it would have skirted several miles to the south of the village. As it was, at 73 miles per hour forward moving speed, it had lost its mass of violently rotating winds within winds and had broken up into three funnels outside Griffin. Although none of the witnesses in Owensville recalled seeing even one funnel, three were, as mentioned before, verified by ground tracks and by other eyewitnesses before striking the village.

Between just outside Griffin to the east, and about the same distance outside Owensville to the east again and a bit north, a total of 85 farms were completely destroyed. The productive land was robbed of the implements, the people, and the wherewithal to produce upon it. Livestock, machinery, storage and mill and granary structures, all were taken away by the killer winds. It was as if the tornado knew that it was about to heap devastation upon the land that was not soon to recover; the Great Depression of 1929, just four years later, came in to finish the job. It was years before the farmers in the Hoosier state were able to adequately find their footing in the farming industry again.

Leaving the charming little town of Owensville in the dust and debris of its sweeping terror, the Tri-State Tornado churned madly on, maintaining it record-setting 73 miles per hour forward speed. Behind it lay six dead in the Owensville area, 47 injured, and $600,000 in damages.

Ahead of it lay the last stop—Princeton, Indiana.

The time was approaching 4:18; the death machine was winding down, but it was not finished.

Chapter Fifteen

Princeton, Indiana

At the end of the 20th century, Princeton, Indiana, was one of those towns experiencing the inevitable growing pains inherent with success.

Gibson County, which held Princeton as well as Owensville and several other communities of varying sizes, was settled in 1789, on the south bank of the Patoka River at a place later known as Severns' Bridge, this by a man originally from Wales by the name of John Severns. Gibson County was organized on March 9, 1813, and was named for General John Gibson, a hero in the French and Indian War.

In 1814, Princeton was chosen as the county seat. The first courthouse, a one-story brick building, was constructed on the courthouse square, a beautiful spot in the center of a busy downtown, in that same year. In 1843, it was replaced by a larger brick building, and again in 1884. Such careful consideration of the county landmark brought attention to the fine work, and in 1984, after one hundred years of upkeep of the fine courthouse that had been established, what became the centerpiece of the town was added to the National Historic Register. It was a stunning work of architecture, in the Romanesque Revival style, and had withstood many a stormy season.

Princeton, on the whole, suffered from the growing pains that befell any successful city at the end of the twentieth century. There were street blockages and bottlenecks that hadn't occurred twenty years before; there was the stretching out of the mecca-like shopping plazas of K-Mart and the competing Wal-Mart just across the highway from each other. The major employer of the area, Toyota, had located its sprawling offices and factories on the southern edge of town (between Princeton and Fort Branch) where once only corn was produced. Pickup trucks, sport-utility vehicles, precision motors and electronic components were just a few of the products being manufactured at the Toyota location. It was a dream to be employed there, being well-paid and well-benefited as a result.

The Gibson Generating Station, formerly known in the 70s as Public Service Indiana, then becoming Cinergy in the 90s (later to become Duke

Energy), was another prime employer of southwestern Indiana residents and Southern Illinoisans alike. It was the largest generating station in Indiana. Electricity from the plant's five coal-fired generating units was used by Hoosiers in 69 of the state's 92 counties. Smokestacks from the plant were a source of amazement for many children, mine included. It was common for my younger daughter to spout, "Momma, that place is making clouds!" as we passed by.

Coal was plentiful beneath Gibson County; Princeton itself sat atop veins of coal and of the tunnels underground that were used to dig them out. It was often said that Princeton was nothing if it wasn't sitting atop a network of tunnels, all dug out from underneath and waiting for the miners to come back and finish the job. But, like the Illinois coal from Franklin and Williamson counties, the bituminous black fossil fuel that was plentiful beneath Indiana's surface, was laden with the substance that created illegal levels of sulfuric emissions when it burned. The Environmental Protection Agency had its say in the early 90s, and at that time only a bit of the coal that lay beneath the surface was mined, enough to keep the generators of Cinergy going. Cinergy had the distinction of having 'scrubbers' in their filtering system for emissions from the burning coal to thwart high levels of sulfur from reaching the air the Hoosiers breathed. Cinergy, being what it was, could afford it. So besides the generating plant, there was enough coal mined to ship to other areas of the country which had factories that were also able to filter the sulfur out of the smoke more efficiently (albeit expensively) than what was available in Illinois and other locations in Indiana.

The locals were kind, easygoing folks, proud of their historical town and county and of the outstanding features they held. They remembered the Tri-State Tornado as well as any of those in the other larger towns stricken, like Murphysboro or West Frankfort, but they didn't carry the bitterness the former did, it seemed. Like Illinois towns, Princeton lost major industry and employment to the twister: the Heinz Catsup factory was ripped apart by the winds, and the Illinois Central rail shops, reminiscent of Murphysboro, were effectively destroyed. However, unlike the western Illinois towns, Princeton recovered and went on to achieve relative prosperity. By contrast, with Murphysboro's loss of the railroad offices and shops, and West Frankfort's diminished coal industry, it could fairly be said the tornado lead to the downfall of what could have been, and once were, prominent Illinois towns.

238

But in southwestern Indiana, life went on after the tornado, and Princeton was alive to prove it. So, too, were survivors. While only two percent of the population of Princeton was killed or injured in the storm, the numbers were no less important: in a town of almost 16,000 during the mid-1920s, the 45 dead and the 152 injured were as staggering a toll today as they were back then.

Even though many knew about the destructive storm and the havoc it wreaked upon the unsuspecting town, there were fewer in Princeton to speak with about it than anywhere else. It might have been that for those who remained who were survivors, it just wasn't something they wanted to talk about. It might just have been too traumatic for any of those to speak up and reveal secret stories about how, as children, they lived through what was very likely Princeton's worst nightmare. Or it may have been that, unlike any other more rural or village-like setting, there just weren't that many people around who remembered the catastrophic event.

That may have been why one of only two stories I was able to secure on my many visits to Princeton was from a lovely lady who worked at both the area Chamber of Commerce and at managing her own secretarial services. Arline Riley was that woman, and although it wasn't her own, she had a story to tell. Her parents, who lived through the tornado, had passed the information to her over the years, as she was born in the time following the great storm. Theirs was a story of such tragedy, such a level of poignancy that it could almost have stood within Princeton's chapter on its own. Arline's was the first of the two remarkable stories chosen to represent the horror that the date March 18, 1925, brought to the residents of this sleepy, historical southwestern Indiana town.

"Right before she got to the door, the wind blew (it) open and the linoleum went up and hit her in the head. And that's the last she knew"—Arline Riley, not yet born on March 18, 1925

"I wasn't born then!" Arline Riley insisted on that unseasonably warm winter night in December 1999 when I finally got to talk to her. "I didn't go through the storm, so I'm not what you could call a 'survivor.'"

Like Charles Price before her in Griffin, Arline was on this earth by the grace of God, her parents, Mr. and Mrs. Charles Whitten, having outlived the

Tri-State Tornado thus earning themselves the badge of 'survivors.' But by turns, Arline Riley, too, was a survivor of Princeton's darkest day, when the afternoon air, warmed from a muggy mid-March morning, grew still and ominous right before turning vicious and deadly at about 4:15 that Wednesday the 18th. Aside from her reluctance to claim herself as such, she was a survivor in every sense of the word.

She told the tale as it was so often related to her throughout the years, and it was as if she had been there to witness it herself. Perhaps it was a reflection of the depth of emotion in her mother's words that allowed Arline to tell it so well. Her mother, long gone, was with her in spirit through the telling. The intensity of it was too powerful to ever let that ghost pass.

"I came along late," Arline said. "Mom was 40, and Dad was 46 when I was born."

Not that that was an odd occurrence; in fact, for people of the storm, it was quite frequent no matter what the age of the parents, children were born in the year following the ordeal. For some, it seemed to be a celebration of life and the continuation of it. For others, it was a replacement of life that was viciously ripped from them in the horrific minutes of the afternoon of March 18, 1925.

For Mrs. Charles Whitten, the Wednesday afternoon that threatened to soak her mid-week laundry on the clothesline appeared little more than that—a threat. She could not know that by 4:15 p.m. that day, so much would be lost, taken from her and from her family forever.

> *"Marybelle said Grandma and Mother had so much sand and mud on them, the neighborhood women came to clean them up before the funeral"—Ada McClurkin, age 5 on March 18, 1925*

Two pretty ladies sat in their living room late on a December winter day 1999, rain pouring on and all around their pleasant little home in Princeton proper. Warm lights glowed throughout the interior, warding off any ghosts of the dark day in 1925 still lingering there.

Apparitions couldn't help but linger around the McClurkin household. These two precious ladies, Harriet, who was 80, and Ada, 79, in December 1999, had been inseparable since toddlerhood, and they had the photos, the stories, and the sisterly love to prove it. Within the photos and the stories were woven the fabric of the ghosts themselves. And, even at so young an age as the sisters were when the storm struck their Princeton farm, the vivid memories were still with them years later. Perhaps, in and of themselves, the sisters created the ghosts, keeping them active, waiting, just outside the home, in the rain, their ghostly hands cupped against the glass in the window— watching those within telling their story. Its poignancy ran a close parallel to that of Arline Riley's. The big difference was that Arline hadn't come along yet when the storm left its violent imprint on the lives of those in Princeton. These ladies had lived through it.

"We lived just west of King's Mine," began Ada.

King's Mine, she said, was due west of the cloverleaf highway, which was formed by Indiana 65 and 41 intersecting on the west side of Princeton. The mine was no longer active and had not been for a number of years, probably, said Ada, since the early 1980s.

"It was a deep mine," Ada explained. "Tunnels all under Princeton. It was too bad it shut down. One of our classmates who had worked there took us on a tour for our twenty-fifth high school reunion," she said with a measure of delight. "I remember him walking us through for such a long time, then stopping and telling me that we were standing right under where our farm was."

In 1999, the McClurkin farm no longer stood on its 1925 location. It was once a viable and active farm, a source of livelihood for Arch McClurkin, his wife Mary, son Morton and two darling daughters, who were only a little more than a year apart in age. As young children, toddlers really, the sisters remembered their family being well off enough to own a car. They also remembered having animals of all kinds, and the crops that those animals helped their father farm for a living.

Then, when the girls were still babies, their father was doctoring a mule. The ornery and offending animal kicked him hard in the stomach. Within a matter of hours, Arch McClurkin was dead from severe internal injuries, and

Mary McClurkin was a widow. It was the type of accidental farm death that was infrequent but happened. Mary McClurkin was the type of farm wife who accepted the hardship and moved on, choosing not to let it be a hardship if she could avoid it, devastated though she was.

She took her three small children and moved to a smaller house on the

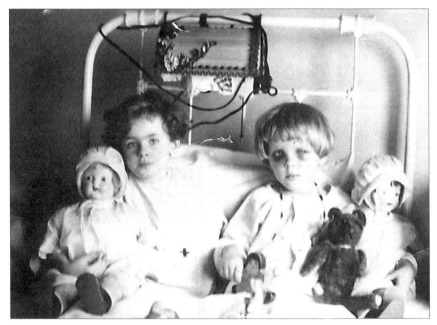

Harriet and Ada McClurkin during their hospital stay in Evansville, March 1925.

same property, and chose to raise them with the help of her mother, Mrs. George Miller, herself a widow, and sister Kate Miller. Arch's sister, Edna McClurkin Armstrong (and an aunt of Owensville's James Richard Armstrong), moved with her family to the big farmhouse after Mary and her children vacated it.

The tragedy for the McClurkin family was not over with the passing of Arch McClurkin. A little more than two years later, another would come along, affecting their lives irretrievably and forever.

*   *   *

Arline Whitten Riley's mother, Anna; father, Charles; and two of her brothers, Wallace, 4, and Charles, 16, were in Princeton for Bargain Day. Another brother, Edgar, 13, was at St. Joseph's on the east end of town, having music lessons.

Bargain Day was a Wednesday occurrence designed to bring customers in from all around Gibson County for special sales in the downtown stores. Drawings and door prizes were held within many of the businesses. It was

The McClurkin sisters seated in their home December 1999, recreating the post the Evansville paper ran of them in a local hospital, with Harriet on the left and Ada on the right.

spring and it was time to be thinking of new items of clothing to match the southwestern Indiana climate, which was notorious for changing rapidly and without much notice.

March 18 was a prime example. The day had dawned damp and chilly; many had stoked the fires or furnaces within their homes. But as the day went on, the temperature had risen. The skies had remained rather overcast, and an occasional ray of sun would break through, but otherwise, sprinkles of rain had continued on and off for the better part of the afternoon.

Now, said Arline, her father Charles was gazing with anticipation at a darkening sky to the south/southwest, and growing concerned. The time approached 4 p.m.

"There they were, all in town, and a storm was coming up," said Arline. "So Dad told Mom to take Wallace with her and get on the motor line, that they called the 'Traction Line,' that ran on the tracks between Vincennes and Evansville, and get home."

The older brother, who had been named Charles after ten generations of

Charles Whittens, had opted to stay with Dad and ride out the storm downtown. In the end, it was fortuitous for them that they did. Brother Edgar, felt the elder Charles Whitten, would likely be safe at the parochial school where his music lesson was to go on for at least another half-hour; the buildings of the school were well-constructed and he would be sheltered by the oncoming rain, which looked as if it could be copious, by the size and color of the approaching cloud. By the time the lesson would be done, the storm would have blown over and Edgar could take the Traction Line himself, as he had his own change.

The 'Traction Line' was a mode of transportation, like a trolley, that ran the city's tracks and would carry people from the outskirts of town to downtown, with many stops in town on the way.

"It was what most people took to get around," Arline claimed her parents told her. "It was a few passenger cars that ran a lot from McCall's Summit, to the west, and Baldwin Heights, to the east, taking people home or into town."

It was a convenient mode of travel into the business district from most neighborhoods. The Whittens lived in the Baldwin Heights area, tract housing set up for those who worked in the local industries of coal mining and factory employment, and took the Traction Line frequently.

The factory concept the likes of Toyota, by the time it came to Princeton in the late 90s, was obviously not a new one. The H. J. Heinz plant, where many of Princeton's residents were employed making Heinz products such as catsup and tomato sauces, was across the main street (Indiana Highway 41) separating Baldwin Heights from the rest of the town. The catsup factory was an imposing structure, three floors, spread out over most of a city block. It kept the Traction Line running from the factory/Baldwin Heights area into town a busy one.

Charles Whitten directed his wife Anna to take their four-year-old on home via the Traction Line, then wait there for him and their 16-year-old son to arrive by foot, should the weather turn particularly bad. The two Whittens boarded the Traction Line and paid their fare; Charles and his elder son started walking.

"Money was tight," Arline explained, "so Dad could only afford to send Mom and Wallace on the train."

*   *   *

Even though she was only five years old at the time, Harriet McClurkin

What remained of the Heinz catsup factory in the Baldwin Heights section of Princeton after the storm. The Whitten men took refuge next to the building which proceeded to collapse around them.

remembered March 18, 1925 to be a "cloudy day."

Her recollection from there extended to one of a five-year-old's favorite pastimes—eating.

"Now, I loved rice, and we were having rice for dinner that night," she said.

She admitted ignoring the usual time set for dinner in the McClurkin household and to going on out to the dining room to have a bowl of rice before supper. Harriet found it amusing, both the fact that she loved rice so much and the fact that, after so many years, she remembered so distinctly sneaking some before dinner on that fateful night, and giggled about it before she continued.

"I hurried and finished because the wind was blowing so hard," she said. "I don't remember anything else except asking Mother, 'Is it going to blow our house away?'

"Then a puff of smoke came out the stove from the flue. Flames came out around the door, too."

That was all either sister could recall in the moments before the monster storm struck their childhood home.

*   *   *

In the downtown area of Princeton, the wind was whispering about a violent, blackened cloud that could be seen stalking the southwest part of the town.

Charles Whitten's thoughts were toward his wife and small son, who, he felt, had possibly escaped the wrath of what was coming by getting to their home on time via the Traction Line.

At least, that's what he *thought* they had done.

His next instinct was one of self-preservation. He took his son, Charles and headed out on foot, according to Arline Whitten Riley.

"The overpass that's now down there in the Baldwin Heights area *was* highway 41 back then," Arline said. "That's what's the business 41 now, that will take you off Indiana 41 and into the downtown area. That was what they took when they went walking home."

The highway passed right alongside the Heinz factory; the overpass, which rose in a giant hump over the railroad that bisected Princeton, was the marker for the exit into Baldwin Heights. It was to that location the Whitten men and the bloodthirsty storm arrived at the same time.

The men went running toward the catsup factory, which was full of workers busy on their second shift of the day.

"They got up against one of the brick walls to protect themselves," said Arline.

The storm, the girth of which was not quite as wide as it had been in previous stricken communities but was just as intense, slammed into the factory like a maddened, careening locomotive. It sheared off the top floor, spraying bricks and mortar in every direction. Heavy bricks and fragments of bricks became tiny missiles raining down from the sky. What walls were not demolished and sent flying collapsed upon those working inside, many, up to that point, oblivious of the storm. Several were crushed in seconds by the weight of the jagged rubble.

The Whitten men backed up against one of the brick walls in an effort to protect themselves, Arline said. And unlike many within the building, they were preserved by sheer good fortune.

"The storm lifted the wall up and over them," she said, "and blew it apart."

The bricks that made up the wall were among the many simply whisked

The devasted Baldwin Heights on the south side of Princeton, Indiana.

away in the rapidly twisting winds. The wall above the Whitten men was there one minute, gone the next.

"It didn't touch them," Arline said.

On the other side of Highway 41, however, it was as if the deadly winds had other plans for the remaining two Whittens.

*   *   *

"Aunt Kate (Miller) and Grandma were staying there in the little house with us that day," said Harriet McClurkin, "and Grandma had gone to the kitchen to get supper."

The girls' grandmother was named Harriet Miller; the elder of the two little girls had been named after her. This namesake, five-year-old Harriet McClurkin, said she had gone out to the hallway between the living room and dining room; the hallway had a staircase ascending, upon entry, into the home.

Into the little hallway came Aunt Kate, close behind Harriet, worried about the tiny girl being out there alone.

"That was when the outside door blew open," Harriet said. "So Aunt Kate went out to close the door…and the windows blew out."

Ada McClurkin claimed she didn't remember the windows blowing out. She did, however, remember the puff of smoke sister Harriet recalled seeing when the first tremendous gust of wind struck the farmhouse. Ada acknowledged the windows were shattering at that point, and Harriet went on with her account.

"When the windows blew out, the wind was so bad that it came through the door Aunt Kate was trying to close and it blew her back onto the stairway," Harriet continued.

The stairway of the little home was constructed of wood, and was being torn apart by the incredible velocity of the winds. Kate Miller was thrown with such brutal force that she landed primarily on the left side of her body. The resulting injuries were horrific. Not only was her left forearm broken in at least one place, but something either threw her across the ragged edges of the disintegrating stairway, or an unknown object viciously sliced across her torso. Whatever the case, Kate Miller's left breast was almost severed from her body as a result.

The worst injury occurred when Kate landed against the stairway itself. Large planks and boards, splintering in the high winds, bared their jagged nails.

"A board with four nails on it was driven into her back," said Harriet.

She was uncertain as to whether the board came away of its own accord in the winds, or Kate removed it herself. At any rate, the object was embedded for a period of time—long enough to pose a considerable health hazard for Kate Miller.

*   *   *

Arline Riley related what had happened at the Whitten home located deeper in Princeton when her mother arrived from the traction line with little Wallace in tow.

"Mom had enough time to get out and start taking in the laundry off the clothes line," Arline said.

Anna Whitten very likely had no concept of the level of sheer danger in the approaching storm; or perhaps, she was hoping, by performing the usual tasks expected of her prior to an average rainstorm, she was averting the possibility the tempest could become so wicked.

However, "The storm kept gathering," said Arline.

Her mother caught sight of it in the southwest, and, observing the amount of wreckage boiling in the air around it, she knew it was obliterating the other side of town.

Panicked, she hastily picked the clothes off the line and pulled them into the house, where Wallace sat in the kitchen, waiting. Anna knew the storm was dangerous, reported her daughter. She was simply at a loss as to what to do as

Scene of devastation at the railyards in Princeton, Indiana, following the tornado.

the killer winds raced toward their home.

Common sense dictated her next move.

"Back then nobody locked their houses," Arline said of a simpler time in human history when even the worst of neighborhoods were relatively safe against theft or intrusion. "So when the storm blew the front door open, Mom had no way to lock it.

"We used to place a knife in the door lock to lock it," Arline explained.

The flat knife blade kept the tumblers inside the mechanism that allowed the door's knob to turn, effectively placing a lock on the knob.

"So Mom told Wallace, 'Go get Mommy a case knife out of the kitchen,'" Arline said.

Little Wallace went from the living room, through the dining room and presumably on to the kitchen. But either he did not continue on into the kitchen and stopped short in the dining room, or went to the kitchen but returned promptly—without the case knife.

"He got behind a coal stove and he said, 'I 'fraid, Mommy! I 'fraid!'" Arline said.

Wallace stayed in his position behind the coal-burning stove as his mother left the front door, went to the kitchen, and got a case knife. She didn't seem to be alarmed or annoyed that the little boy had gotten himself out of harm's way. All she felt she could do to provide protection was to simply keep the front door to the house shut. It would appear the young mother simply wanted to keep the hands of the killer wind from wrapping its fingers around them both by locking it out of the house.

However, that was not to be the case.

"Right before she got to the door with the knife, the wind blew it open and it blew the linoleum rug up and hit her in the head. And that's the last she knew." said Arline

*   *   *

On the McClurkin farm southwest of town, two little girls were rousing themselves up from the rubble that was once their home.

Harriet McClurkin revealed the first moments she remembered following the assault.

"I know Ada was trying to wake me up. She was on the ground there with me," Harriet said.

Ada spoke of her initial shock and the response she had during it. "I was awake first. I was trying to get mother up, too."

The imagination can only take over at that point, in an attempt to understand what it must have been like for two little girls, ages four and five, to come back to consciousness in a pile of ruins that had been their home, and find their mother lying still nearby. The attempt to rouse her was unsuccessful, they reported, and, full of fear, they managed to get each other standing and spotted their Aunt Kate lying in a nearby ditch.

"We got her up," Harriet said.

The woman was gravely injured; but the only thoughts on the girls' minds were to get their mother up and about and get to some kind of shelter, where they could be still and hopefully calm down from the terror that had invaded and taken all comfort from them.

"We looked around and saw a building to the east, and I asked should we go there, and Ada said no," Harriet recalled.

The girls' uncle, Art Armstrong, who lived nearby, had gone to the Milton Woods residence north of the McClurkin home.

"His daughter Nancy and our brother Mortin were coming home on the school bus, and Uncle Art went to get them when he saw the bad weather coming," Harriet said. "The bus driver told them to get out and into the ditch. The driver didn't. The kids weren't hurt, but the driver was."

Harriet was unsure of what exactly happened to the bus in the storm, but

Scene after the tornado roared through Princeton, Indiana, on March 18, 1925.

she recalled that Uncle Art came to claim the kids and then said he'd have to go see what had happened at his place.

"On the way, he stopped at our house," Harriet said. She reflected thoughtfully for a moment, and said, "He couldn't have been too long—he had to go take care of his kids."

The first thing he took care of at the McClurkin residence was the one thing the girls had feared from the moment they both revived.

"We didn't know Mother was dead," Harriet said. But by the tone in her voice, they knew that Uncle Art *did* know.

"He looked around and said, 'We can't do anything else here,'" said Harriet.

She paused, thought about it quietly, and then offered in a serene voice of resignation, "I remember the horse, Dot, running around after the storm."

"So do I," Ada said with the same serenity and resignation.

*   *   *

"My dad was down there, bricks all over and around him, and the storm had subsided," Arline Riley said, "and he started walking two blocks up the hill and two blocks south, that was all the farther from the catsup factory he needed to go to get home."

However, Charles Whitten discovered as he exited the factory area, it was going to be difficult to get home without landmarks.

"Everything was gone," Arline said about the south end of town. "It wasn't easy to find where you're going…when the landmarks are gone, you don't know."

The elder Charles Whitten and his son Charles struggled across a field of debris, timber and bricks, inching their way toward the little house in Baldwin Heights. All around them, other homes were in various states of damage ranging from lifted off their foundations to collapsed to simply obliterated. Trash, clothing, household items, dishes, everyday belongings of people they knew and others they likely did not lined the way they traveled in order to get home.

It was as if time were trying to spare them the sight that they were about to behold.

"By the time they got there," said Arline, "the foundation and coal stove were all that was left."

Arline paused before she said, "Mother was sitting on the floor, which was elevated by cinder blocks and somehow didn't get blown away. Her legs were dangling over the side of the foundation.

"She was holding baby Wallace. He was dead."

Arline gave another pause; it didn't appear to matter that she was not even a twinkling in her mother's eye at the time the abomination had happened. She was living it; she was Anna Whitten's daughter who came along after the storm, and she had heard it all and had seen it through her mother's eyes and felt it through her mother's arms.

Those arms were wrapped tightly around the little broken body; the boy's own arms hung limply down and across Anna's lap. His mother's arms were those long-practiced in comfort and love and she seemed to hold all comfort and love in that single embrace. But the one final embrace was too late. Though no one ever found out the specific details of Wallace's death, it was often relayed to Arline that it was suspected the little boy died from a brick hitting him in the head, likely when the fireplace above him collapsed in the violent winds. Whatever the case, the boy's life was extinguished, and his mother's awareness of the fact, the same.

"He was dead and she was holding him. She was awake but didn't know anything."

Anna's legs were swinging back and forth dangling over the foundation as if she were a child herself, hanging her feet over the water out on a dock. According to reports, she held her baby Wallace in an embrace as if he were four months old instead of four years old.

"They carried her to where they could get transportation to the hospital; nothing could get through the streets down to the house," said Arline.

"They buried Wallace before she ever knew he was gone."

*   *   *

Uncle Art Armstrong was able to get to the McClurkin girls and their Aunt Kate Miller and begin walking with them over to Aunt Edna Armstrong's, on the same property as their house that had been completely blown away in the storm. It was a bit of a struggle, but it was going to be necessary, as their wounds were serious enough to necessitate getting all three to medical care as quickly as possible.

"Harriet had a broken left arm," said Ada of her sister, "and my eyes were hurt."

"And Aunt Kate was all banged up," Harriet reminded her.

The piece of wood with four nails which had been driven into Kate Miller's back had been removed, but the injury hindered the woman, along with her other injuries of a broken arm and horribly torn breast. They walked along the short distance to Edna Armstrong's home.

"As we were walking with Uncle Art, he and Harriet were ahead of Aunt Kate and me," said Ada. "Uncle Art was hard of hearing. So Aunt Kate said 'call to Harriet and have her tell him to slow down; I can't keep up with them.'

"I remember seeing Harriet turn around when I yelled at her and that's the last thing I saw—my vision went at that point," Ada said. "It didn't panic me…I just don't remember anything else."

Before her vision went, Ada clearly remembered the sunset, and the fact that, following the storm, "It was so much lighter because we could see as we were walking across the field.

"Somebody came, I don't know who, and took us to George Ward's house and bathed us," Ada said. "They didn't have any girls there at the Ward's, just boys, so we had overalls on when Colvin's ambulance service came to get us and take us to the hospital in Princeton."

When the girls got there, they were amazed at the number of people with which the hospital was filled.

"They laid us out in the hall on the floor, me, Ada and Kate," said Harriet.

The doctors, somewhat overwhelmed by the amount and extent of the injuries coming in to the hospital, were treating as many as they could, as quickly as they could.

"Everything was full. I remember two people, doctors, got hold of my arm and pulled it and set it right there," Harriet said, the pain of this medical treatment still evident in her words.

Ada's eyes had been damaged by something in the winds that she did not remember encountering. The left eye was in particularly bad shape, banged and bruised, swollen completely shut and extended beyond the socket with the swelling.

"Dr. Rhoads, an eye specialist in town, wanted to take the left eye out," she said, "but they didn't."

On being temporarily blind, Ada remarked, "I kinda accepted it in the hospital. They kept my eyes all bandaged and finally lanced above the left eye to let the pus out of it."

Her injury remained a mystery, even to those in the medical field, as to how it was inflicted.

"My nose wasn't hurt," she noted, "the back of my head wasn't hurt; just my eyes."

Harriet commented, "I had no other injury, just the broken arm."

"I've often wondered if they kept her in the hospital just to keep me there," said Ada, and the sisters laughed.

But their Aunt Kate remained in the hospital, and her stay as well as her wounds weren't so funny. Sadly, the sisters recalled that Aunt Kate had been a victim of her own injuries in more ways than one. Due to the extent of her puncture wounds from the embedded nails and the torn breast, the doctors overlooked the fact that Kate's arm was broken. It was never set, and the result of this was that Kate always had a slight bend in her left forearm, and she could never effectively rotate it normally.

"They didn't set it because they didn't think she was going to live," Harriet explained. "They had other things to worry about that were wrong with her."

"Kate always said she was the last one *in* the hospital and the last one *out*," Ada said. "She went in with us on the ambulance. They didn't have any more room at the hospital after they admitted her. That's why she was the 'last.'"

Kate's injuries kept her in the hospital for several weeks. The girls stayed for an approximate total of three weeks. They missed the burial of their mother, and of their grandmother.

"I don't remember finding out about mother," Harriet said about the death of their only remaining parent.

"They had mother and grandma's funeral at the same time," Ada said. "Marybelle (another relative) said grandma and mother had so much sand and mud on them that neighborhood women came to clean them up before they buried them."

Grandma Miller's body was found along a fencerow several yards from the farmhouse.

"It was after Uncle Horace came and got mother," said Harriet. "He said, 'We're gonna have to go back and find grandma tonight.' Uncle Henry (Woods) said, 'You can't. You can't go touch a body after something like this, it's gotta be official people like from the coroner's office or the morgue to touch the body if you find it. We can't go out and look tonight.'

"Uncle Henry was really upset," Harriet said. "He probably knew that wasn't right, what he was telling Uncle Horace. But he wasn't thinking clearly. And anyway, Uncle Horace went back and found grandma. Every bone in her body was broken and her arm was all wrapped up in the fencing. Uncle Horace was glad he found her. He was afraid a wild animal would be getting to her before daybreak."

A photo of the two girls sitting in their hospital bed with their dolls and toys graced the pages of the *Evansville Courier* a couple of weeks later. The two sat rather morosely; it was evident the trauma they had gone through, not only because Harriet's arm was visibly splinted and in a Red Cross sling, and Ada's eyes were bruised and battered, but because they were two orphaned little girls who could not muster a smile for the photo.

Because Aunt Kate was set to raise the girls once she was recovered, the newspaper was forced to run in an article following publication of the photo and many, many inquiries:

"The McClurkin children are not up for adoption. The family will take care of them," the paper wrote in response to the overwhelming number of calls they'd received containing inquiries as to the girls' familial status…may had outright offered to take the girls home with them upon release from the hospital.

*   *   *

Arline Whitten Riley's mother Anna had "many injuries," she reported.

"She had severe head injuries, a concussion, broken ribs. Her leg was torn open between the knee and hip. They said you could see the ligament and stuff inside the gash, working in there," Arline said. "Even after I was aware, she had a gash six inches long.

"You know how they doctored it?" she asked. "They poured alcohol in it and left it.

"All the times this has been retold to me it's been told that way," Arline went on. "All Mom remembers is this, up to getting hit—then waking up in the hospital, where she was for two weeks."

Anna was not blissfully oblivious to her surroundings, however.

"Not all that time was she unconscious," said her daughter, indicating that Anna had time to lie in the hospital bed and wonder about her son, whose funeral she had not attended, and how the horrific events of the day came about so quickly and with deadly finality.

*   *   *

The swiftness at which the storm came upon the area of Princeton was not lost on two little girls, who, years later, testified to the enormity of the tornado and its effects on it victims later in their lives.

"The only thing I ever say," said Harriet McClurkin, "is—you don't know how quick it is. Those people who watch storms coming in, I say, You don't have time. Don't do that."

"From hearing the stories as a kid," Arline Riley concurred, "I liked to never got over storms." And Arline, as pointed out before, was not even a literal survivor of the Tri-State Tornado.

*   *   *

At 4:18 p.m. on March 18, 1925, the storm, exiting out of Princeton, Ind., left its last victim. The Tri-State Tornado looked to be heading toward more densely populated Petersburg, Ind. On its unwavering heading of now N 60 E degrees, at about 16 miles northeast of Princeton, moving forward still at 73 miles per hour, the tornado finally reached the end of its uninterrupted path of 219 miles on ground.

The storm had lost the steam that many experts believed was given it by the jet stream from the upper atmosphere, and was now effectively tearing itself apart, moving into less favorable conditions. Without the sweeping upper level winds to propel it, the supercell was running on its own momentum and

A view of the damage on the rail line and cars in Princeton. The remains of the catsup factory stands in the distance.

found it couldn't support itself. It fell away from the jet stream, slowing and dying a very natural death, nothing like what it had heaped upon three states in three and a half hours, no. This death was in the breath and sigh of winds that almost instantaneously went from the raging and furious to breezy gusts to puff. The funnel diminished from a boiling screaming cylinder to a thin, whiplike cord.

Minutes away from Petersburg, the tail of the sharp, funnel-shaped cloud lifted off the earth, and in the twinkling of an eye, very unceremoniously dissipated while over an unplanted corn field, harmless in an unoccupied, nondescript area of Pike County in southwestern Indiana.

The time was 4:30 p.m.

The horror of Wednesday, March 18, 1925, was over. Left behind in Princeton were 45 dead, 152 injured, and $1.8 million dollars in damage—the third highest dollar amount total of its three-hour killing spree.

The top third of the Heinz Catsup factory was gone. Many of the Central and Eastern Illinois rail shops were obliterated; the roundhouse and shops of the Southern Railway were severely damaged. The south end of the town lay in rubble, and the damage appeared horrific, as if a war had taken place within and surrounding the Baldwin Heights area. However, only two percent of the town's population was dead or injured; as harsh as it sounded, by number,

Princeton was in the minority of body counts that day.

But that statistic could not be told to Anna Whitten, who lost a precious boy, nor to the little McClurkin girls, who lost their mother, the only surviving parent they had, and be expected to be understood. The death totals in the Tri-State Tornado bore this out repeatedly from town to town and village to village, if it bore out nothing else:

One life taken by the storm, to the survivors, was, very simply, one too many.

The front page of the March 19, 1925, edition of the *Princeton Clarion-News*.

Part Two: Aftermath

Chapter Sixteen

Aftermath

They were like ants in that their homes, mounds of sand to an ant, buildings of brick and mortar or wood to a human, had been demolished and their lives upended.

Ants scurry about when a child comes along and steps on their sand mounds, grinds the sand into flatness upon the ground and keeps going, oblivious, most of the time, to what he has done. The ants get to repairing immediately, their single-mindedness admirable in that they can stay so focused and determined to rebuild and continue with their little ant lives. Even if a child finds a little section of ant mounds dotting the sidewalk for several yards, and feels compelled to squash the piles on purpose, the ants continue their diligent work. Nothing sways them; it is innate, animal instinct. It is, however, thought that prompts the child to act, not instinct. Regardless of whether the act is done out of meanness or curiosity, the child who deliberately demolishes the mounds gives thought to the act.

The Tri-State Tornado has been personified on countless occasions, called a demon, a wraith, an 'Act of' or 'the Finger of' God. It has been suggested that many of the areas hit by the storm deserved God's justice upon their community, what with prohibition, wild times, and industry corruption running rampant in their town.

But it can also be fairly said that the Tri-State Tornado of March 18, 1925, was an extreme event, a very large storm that took an odd shape that may historically have happened in the past and might occur again, but maybe not on this exceptional level. In this regard, it is unreasonable to think the damage, the hideous deaths and destruction, were anything purposeful like a child treading on sand mounds of ants. Nevertheless, some were able to start picking up the pieces, many no bigger than piles of sand, and putting the fragments back together again.

Homes were broken, twisted, bent, skewered with parts of other homes, shattered, reduced to rubble, or burning with fires ignited by flames that grew from an overturned coal- or wood-burning stove. Many homes were simply swept away, not a splinter of them ever to be seen again.

Businesses were done the same, their modes of trade, inventory, paperwork, monetary accounts obliterated, damaged beyond repair, or disappeared in the blackness that surrounded the debris field, again, never more to be seen. Specialists in areas of trade, such as rail shop workers and machinists, were killed without passing their trade on to others. Over time, many of the skills became lost with advancing technology and modernization, but the history of the work died with the craftsmen, and, like the homes and businesses, was taken by the storm.

Entire families were wiped out. Many across the damage path, from Missouri to Indiana, had husbands/fathers who were down in the mines, only to emerge to find their wives, sons and daughters dead. Some never remarried, so torn with grief and unable to recover were these. Many families, especially in the Murphysboro/DeSoto area, were left completely childless, and thus their lineage was ended because some, being beyond childbearing age, simply could have no more children. Many children were left not only homeless but also orphaned. Some, as in the case of babies found by couples and raised as their own, never came to know their actual parentage.

The storm moved so swiftly and was so instantaneous, especially in Missouri and Indiana when forward moving speeds reached over 70 miles per hour, that it seemed unreal that so much damage could be caused by something so quick. Survivors were left bewildered in a pile of rubble, or far away from where they were when the storm struck. Wandering the ruins of what was their home or hometown, they seemed dazed, knowing, not to immediately begin the pick-up and get back on their feet, but only to take another breath in order to continue to survive. Within that breath would be the intent to see what could be done, if anything, to take the next step toward recovery, to re-establish normalcy once again.

These are their stories, the aftermath.

Leadana/Annapolis, Missouri
Dead: 4
Injured: 25
Percent of population dead and injured: 1
Property losses, 1925 dollars: $400,000

In the two villages, with populations totaling not quite 700 people, four unfortunate souls lost their lives, 25 were injured. However, according to eyewitnesses, there were less than a dozen buildings standing between the two communities. The twister ripped through Annapolis' downtown section, almost exactly down the middle of the main street where sat not only businesses but residences as well. Those left homeless, by percentage, were practically the entire population.

"Our uncle came as soon as he could, got his wagon and team hooked up and took us to his place," Alice 'Peachy' Jones said.

The day after the storm, according to Peachy, over 200 cars come in from the north to view the damage.

"It was a continual line of cars coming to see and probably as many from the south coming, too," she said.

One might rightly have thought those folks, coming to view a disaster area, would know people were in need, and as a result, bring assistance in the form of food, water, clothing, necessities, anything called for in the face of such devastation. But that wasn't the case in Annapolis or Leadana.

"Nobody brought any help," said Peachy.

But the Red Cross moved in to at least handle living arrangements for those survivors left destitute.

"The Red Cross helped some after that by setting up two tents for a few weeks for us to live in," she said.

The tents were erected on the flooring of the family's home, and there they lived until they were able to clean up the mess the tornado left of their house and begin the rebuilding process. Given the injuries her mother sustained, Peachy felt that her mom was limited in what she could do, but the family nonetheless got by well enough to eventually clear off the land and start anew.

"We moved into another house down the street until our home was rebuilt," Peachy said.

The tents came down, and her father did the best he could with what he had, which wasn't much. Somehow, he had to keep working in the mines and rebuild his own home; the Red Cross' efforts stopped at a certain degree of assistance for the family.

"They helped some," Peachy reiterated of the Red Cross, "but Dad wouldn't rebuild without flues, because of fire hazards, and the Red Cross wouldn't pay for flues. So my brother got a job within a few days, clearing bricks out of the damage areas."

Her brother made good money, as there were a lot of wrecked fireplaces in the Annapolis area. Bricks had rained down all over.

"Dad took excess bricks and they made new flues. The house I live in today is the house we rebuilt, and has the same flue, made with those bricks."

Her mother's injury, which was a 'torn place' in her back, took a very long time to heal, according to Peachy.

"It took months for it to clear up," she said. "We didn't have any money for doctoring. And we didn't get any help from the Red Cross for a doctor, so she just had to deal with it.

"The Red Cross brought cars into town full of clothing, but we couldn't get any," Peachy went on. "That was because our clothing didn't get blown away. And anyway, it was the sizes and types of clothes they had that wouldn't've done us any good. There were some clothes in little boy's sizes, like Archie, and there were some like for my older brother, but there weren't any clothes for growing kids like myself. So none of us kids got any. But we got one pair of overalls for my dad."

The monetary assistance stopped for the Smiths at that point as well.

"It took everything Dad could do," said Peachy, "to rake and scrape to finish the house."

To top it off, the emotional trauma the storm left behind was almost more than many average people could take. Someone who had been weakened by their injuries suffered even more so, as in the case of Peachy's mother, who was terribly frightened of storms as a result of that menacing March day.

"She was scared to death after that. When a storm came up, Mom would gather the kids and go to a neighbor's storm cellar, but Dad wouldn't. She'd say, 'Dad, come on, I'm gonna take the kids.' He'd say, 'No,' and she'd say, 'if you don't I won't either.' And he went.

"She was like that as long as she lived, although it lessened by the time she died at 93. But I remember an insurance man from Poplar Bluff coming to us and laughing at her being so scared. Well, he liked to got killed in that later tornado that came through Missouri. And he came back and apologized to Mom after that."

Biehle, Missouri
Dead: 4
Injured: 11
Percent of population dead and injured: 15
Property losses, 1925 dollars: $45,000

Biehle, like Annapolis before it, lost four of its own to the violent, swirling winds of the storm. The low death toll, however, was no consolation for either community, and even less consolation for the smaller Biehle. With only 150 people claiming residency within the village, they simply had more, by the percentages, to lose.

Julius Hotop, who lived next door to his grandson Bill's tavern and ate lunch and dinner there daily, described frightening results of the storm's devastating winds—frightening because of the force that it must have taken to achieve those results.

"After it was over," Julius said, "my family and the neighbors went out and compared damage. There were roof tiles off some and other buildings were just completely destroyed.

"That storm picked up two boys and they landed in a neighbor's field covered with dirt. They were so dirty there for awhile no one could tell who they were.

"And the woods!" Julius exclaimed. "Oh, you wouldn't believe how the trees were broke off and splintered." He shook his head, thoroughly impressed with the damage inflicted, even after so many years: "Hardly anything was standing."

And yet, a stand of trees made it through the vicious winds of the storm, not broken, not stripped of branches and bark—but they were young, saplings likely, and the part Julius told next only lent to that theory. Many trees survive storms simply by giving within the vortex; twisting with the rotating winds. This was especially true of young trees, as evidenced by taking a sapling, even a tall one, and bending it to the ground. Trees are supple to a certain age and level of resilience.

When they have this resilience, they can twist with the winds as well as bend to straight-line gusts. This is how the phenomenon of "straws driven through trees" in such a violent storm is caused. Trees twist, bark opens up, weaknesses within the structure of the tree are exposed. Whatever object is within the debris field can be captured by the bark or the recesses of the weaknesses—and when the tree snaps back after the winds subside, whatever is in that tree at that moment *stays*.

There was another oddity Julius Hotop experienced in 1999, seventy-four years after the great storm.

"Last year a stand of trees were cut," Julius explained, "and they were twisted, all of 'em, from the inside out."

Other effects of the storm were discovered and dispatched early on in the days and weeks after the Tri-State Tornado.

"My brother and I hauled things in for a week with a team," Julius said of clearing the farm of rubble left behind in the wake of the storm. "I remember there was so much fence wire that was twisted and gnarled up."

Strength of the winds that blew across Biehle, if ever they were questioned, was put to rest with Julius Hotop's next recollection.

"Our neighbors had an eight-foot square cistern rock that lay across the top of their big well," he said. "The wind pushed it almost a foot off the cistern. This is the most amazing thing I ever saw," Julius declared, "because it would almost take a bulldozer to move something that heavy. I don't know how they ever got it back on."

The rural area outside Biehle, beginning at the western county line of Perry County and continuing to the Mississippi River, was stricken over many miles. Four more people perished in the ensuing winds that swept across the hills and forests of the areas of Frohna, Wittenberg, Lixville and Altenburg; 25 were injured.

The twister seemed to slow as it approached and crossed the Mississippi into Illinois.

Gorham, Illinois
Dead: 37
Injured: 170
Percent of population dead and injured: 41
Property losses, 1925 dollars: $150,000

Rosetta Casey Adams, her brother, sister and parents, owed their survival of the Tri-State Tornado to their mother, Edna Casey, who was always on the move. Edna's proclivity for running meant, therefore, that the Casey family not only moved from home to home, renting, but owned a home within the city limits of Gorham to return to if anything should happen. One of Rosetta Casey Adams' childhood homes, the one the family had just vacated the day the storm struck the little river/railroad town of Gorham, had been blown out into the street by the twister. It lay there for several days following the horrific winds.

While many homes within and surrounding Gorham received maximum damage and were reduced to splinters or blown completely away, this little house had been one that had escaped such a death sentence. It had simply tumbled end over end and out into the street.

"Dad got four teams of horses and hooked them up to the understructure of the house," Rosetta said. "And then he put some skids out next to the house, coming off the street, and skidded it across the ditch."

The fortunate aspect of the mess was, after the house had tumbled, it had landed upright out into the street. With the skids and the four teams of horsepower, Ed Casey had managed to slide their little house right back into position. In keeping with the one-hundred-percent destruction that Gorham had experienced, however, the house was unfortunately not completely intact.

"It looked like the wind had sucked the windows and frames and doors and their frames right out of it," Rosetta said.

The sheer number of backyard storm cellars in Gorham, those that are still used in the decades following the storm and those that were in decay or disrepair, are poignant testimony to the impact the devastating storm had on the little town.

Rosetta and many in the village were resolute in their belief that the Tri-State Tornado was the undoing of Gorham. Where the village once possessed a population of over eight hundred, it stood at a mere four hundred at the turn of the twenty-first century. The destruction of businesses and the loss of rail function, even in the minimal weeks it took to get it back to a state of repair, was enough to send Gorham's economy into a downward spiral. The resulting effects of the storm as it stalked further to the northeast was not lost on the residents of Gorham.

"Murphysboro's like Gorham," Rosetta said, "only bigger."

Rosetta Casey Adams at the grave of her beloved aunt Bert Casey in December 1999. Nearly 75 years after the storm Rosetta's voice still filled with grief over the loss of aunt in the great storm.

Murphysboro, Illinois
Dead: 234
Injured: 623
Percent of population dead and injured: 7
Property losses, 1925 dollars: $10,000,000

As the storm swept across the northwest end of Murphysboro, it did more than damage the Mobile and Ohio rail shops, shear the top off the Brown Shoe Factory, and collapse the walls of the Lincoln and Longfellow schools. As with any assault, it left horrifying memories engraved in the hearts and minds of those children who lived through it.

"In 1957," said Mary Belle Melvin, "there was another storm that came through Gorham and Murphysboro, and it had a bad tornado only not so bad as 'The Storm.' But it followed the same track, and did a lot of damage."

Worst of all, it created a lot of fear in the souls of those adults who, as

Remains of Longfellow School in Murphysboro following the tornado.

children more than a quarter of a century earlier, had suffered through the worst disaster their town had ever seen.

"I was driving my husband's truck," Mary Belle said, "and had my year-old son with me and my eleven-year-old daughter. I came under the viaduct under Bridge Street and Laura started screaming. South of the curve was where it was, and it was black to the ground."

Laura, Mary Belle's daughter, knew, from the stories she had heard her mother repeat, exactly what the violently moving whirlwind was.

"We could see the tails whipping out. But I kept saying, we can outrun it; I heard in college that if you try, you can outrun it.

"Up the hill we saw people looking, and it was a bad storm, but the lights were still on. We got home, the kids went to the basement, and the lights went out. I looked out the window, but I didn't pay much attention to what was happening outside. Laura, though, she came up and said she could see red in the skies. I was just saying, 'What's going on here?'"

A little while later Mary Belle's husband came home, and had declared that a tornado had indeed hit Murphysboro—again.

"I asked, 'A tornado hit Murphysboro?' and I meant it. It didn't dawn on me that it hit. Why? It's because, after living through what I have, you don't really think about it. You are in a state of, 'Whatever.'"

FIRST PICTURES OF STORM DISASTER
HERALD CHICAGO EXAMINER

1,000 DEAD, 3,000 HURT LATEST TOLL OF TORNADO

In the Twirkling of an Eye, Murphysboro Was No More

The *Chicago Herald-Examiner* ran some of the first pictures of the devastation from Murphysboro.

The state of 'whatever' was no doubt shock, and seeing such a thing as a tornado snaking across the town that was torn apart by one only twenty-two years earlier, to a survivor, was enough to send her back into that condition, quite without her knowing it.

"But if I didn't know what that was, why did I tell Laura that I heard in college you could outrun it?" Mary Belle asked. "We went through all that and didn't really think about it hitting us."

Her frightening account stems from the experience she had as a child of eight, when, after the 1925 storm had damaged Murphysboro so gravely, she and her family went to stay in a church. That church was near enough to the devastated areas and the fire-wracked locations for young Mary Belle to hear the agonizing screams of those being burned alive. Many people, in buildings that had basements, had run down to them when the storm threatened on the horizon. As a result, the buildings collapsed upon them, and coal- or wood-burning stoves came tumbling down, creating an inferno that engulfed the structures—and turned basements into flaming ovens that slowly consumed those who had sought refuge there.

Scenes of devastation like this one on the west side of Murphysboro presented a daunting cleanup task for the survivors.

"I'm not afraid of storms," Mary Belle insisted. "But I don't go to the basement where everything will fall in on me."

The fires were the second disaster to hit Murphysboro in as many hours when the storm came through.

But perhaps a third wave of disaster, one created by mankind himself, struck in the days following the wind and the devouring blaze, prompting measures to be taken by citizen soldiers in response.

"There were looters. And that meant that National Guard came in," Bernell Howard said, "and that sealed off the town. If you didn't have any ID or a pass, you couldn't come in or leave. They got here the next morning, after the storm hit."

For many, that was too long a stretch without some form of control to monitor the activities of those not injured, opportunists taking advantage of those who were injured.

Bonita Threlkeld, of Hurst, remembered her father, John Brandt, being in the National Guard from Carbondale. Although Bonita wasn't born until twelve years after the disaster, her father's words came back to her and she recalled what he'd told her and the family for so many years.

"Dad was with the Major," Bonita began, "and some other guys,

The front page of Murphysboro's newspaper, *The Daily Independent*, two days after the tornado on March 20, 1925.

monitoring one area, when he saw the Major unsnap his holster and tell a man to stop!"

The man was bent over a dead body that had not been removed from the rubble in the twenty-four hours following the storm. In his hand was the corpse's hand, and the man, a looter, was ready to remove a ring from the finger of the dead body. When the Major shouted for him to stop, he froze.

"Then the Major made this guy empty his pockets, and fingers came out with rings all over them. See, the dead hadn't been dead long enough and the joints on some of them were swollen—so he'd just cut off the fingers—with either a pocket knife or by busting a joint and breaking the finger off, all to get the rings."

The reaction by the Major of the Carbondale National Guard was swift and final.

"The Major shot him on the spot. That was the standing order. If they were caught bent over a body and were mutilating it, shoot them."

It was, according to the military mindset and those who supported it, not out of line at all as a show of force to get the looting stopped.

"Dad said by the blood and dismemberment, you found yourself able to shoot, because what they were doing was inhuman," Bonita said.

Bonita posed the question that must have been on the minds of not only the Guardsmen, but on those who were looting as well, regardless of how they felt about their crime.

"How could you know they were dead?" she asked rhetorically. "They could've been unconscious is all."

The death toll in Murphysboro exceeded that of any other location, even though comparatively, the devastated area was relatively small. Deaths mounted in the highly populated industrial area, where not only were the workers housed, but they were busy at their occupations when the storm swept down from bruised skies at about 2:30 p.m.

"Dad said bodies were coming in so fast they were using the bottling plant as a morgue, stacking them up," said Bonita.

The bottling plant, Stecker's Brewery, (which later became apartments), was a good place to take them; prohibition had shut down production of beer, which was what the plant bottled in Murphysboro, and it was just an empty shell by the time the storm struck and created a need to place so many dead within its walls.

Bernell Howard elaborated on the matter.

"The Elks and Eagles buildings were being used for the injured; the hospital was just a small building and they didn't have the facilities nor the room. The present courthouse was rebuilt at the time, so the former was used as a holding facility for the looters who were caught and not shot."

The Red Cross was the relief agency of the day that took praise for their mobilization actions but condemnation from many of the survivors.

"My Grandma was definitely against the Red Cross, because they didn't help us," said Margaret Russell. "She just didn't get any help from them."

Bernell Howard said, "The Red Cross came in and several church organizations came in and gave aid. But now, out between 20th and 21st streets, on Spruce Street, there's a two-story brick house that was rebuilt by the owner, and the Red Cross wouldn't give him any help, so he put up a billboard criticizing them."

Bernell pointed out that the National Guard brought sixteen by sixteen-foot pyramid tents. Some survivors lived in them up to a year before they got their homes rebuilt.

The Daily Independent—Extra

Thursday, March 19, 1925

DEATH LIST

Storm and fire dead 125 here at 2 p.m. More bodies reaching morgues hourly. Hundred injured. Many seriously.

DeSoto dead reported 118. School collapsed at 2:30. Town flat.

Gorham destroyed. Countryside southwest hard hit. Carbondale net hit. Country safe east of river. Homes flat north of city. Ava, Ora, Elkville, Vergennes reported safe. Medical aid, supplies rushing in.

Italian woman, 4 children	Dorris Stevenson	Ivan Lipe
Mrs. J.W. Gibson	Jerry Callahan's child	Bernard Sheley
J.W. Mifflin	Albert Callihan's child	Mrs. L.E. File
Frances Hammer	Ernie A. Hinchcliff	Minnie Beck
Unidentified woman at high school	Robert Stevenson	B.A. Orland
Joe Moore	1 unidentified girl	E.C. Harris' boy
Loris Miller	Joe Correnti's child	Octavia Trembly
David Ellis	Helen May Cook (child)	Evelyn Boston
Sam Kereens	Columbus Pierson's child	Edna Hays
Arthur Duncan	J.G. Harris	Robert McCord's child
John Hammerhehl and brother Ben	Campbell Lipe	Mr. And Mrs. O.S. Silvey
Mary Davis	H.S. Coontz	---- Gregory
John Swafford	1 unidentified girl	Earl Russell
Frank Baroni	D.E. Darby	John J. Brewer
Mrs. Ardell Spangler and child	Dr. Forshee's son-in-law	3 unidentified children
Mrs. Louis Miller	Mrs. Mary Brandon	Wm. Spurlozzie
3 unidentified children	1 unidentified woman	Dolph Isom, colored
Robert Piltz	Mrs. Geo. Berger	Mrs. Jones, colored
Luella Piltz	1 unidentified man	Mrs. Kelly, colored
Tressy Schmallerbeger	Mrs. Mart Halliday	1 unidentified man
Clara Bailey	S.M. Haney, Meridian, Miss.	Major Verbal and wife
Dr. L.R. Wayman's son	J.A. Jones	Mary Mainard
1 unidentified girl	W.E. Neal	1 Stiver children
3 dead in Tower Grove settlement	Ernest Hardwig	3 unidentified women ---- Clements
August Hassebrook	Ben McAllister	Herbert Lennington
Alworth Gregory	Mrs. James Fielding	Mrs. Sam Rodman
Helen Bowerman	Son of James Fielding	Mrs. Slater, colored
Mrs. George Baker	Claud Lipe	Sister of Necie Coffer
John DeWitt	Jerry N. Mifflin	Joe Henry
E.J. Bjik	---- Willis' (child)	Mrs. Anna Loy
Baker's child	Joe Baroni	Chas. Loy
J.G. Andrews	Albert Nausley, son of Ray	Thomas Loy
John Berra at C'dale	Unidentified man	Earl Varner
Mrs. Jas Martin	---- Roberts child	2 unidentified women
Willis Cochran	Dan Baucher (not D.L.)	unidentified girl
Henrietta Ditzler	Edna Imogene Boucher	Louis Gouldoni
Hassebrock's child	Mrs. ---- Nolte	Mrs. Mildred White
1 unidentified child	Mrs. Wallace Blacklock	Foster Byown

Mary Belle Melvin recalled tents being set up by the relief agency.

"Well, for some, the Red Cross helped and for some they did nothing," she said. "I know my dad got a letter from the Red Cross: it said he was young, he could rebuild."

Laura Marie Miller remembered the Red Cross being present when school started back up in the weeks following the devastation.

"I do remember that," she said, "the Red Cross gave milk at school to

Murphysboro Dead – Continued

4 p.m.

Partial list of hitherto
unreported dead at auxiliary morgue
of Raberts & Sons, in garage at
extremity of Tenth street.

John Schirea and 4 children
James Martin
Unidentified girl baby
John Brewer
Mrs. Spellazzo
Aldencliff Isom
Mrs. ---- Jones

Phoebe Pelley
Alice Brown
Unidentified white baby
Sadia Strattan
Mrs. Mollie Higgins
2 Stribait boys
---- Clements
Jimmy Hays' baby
Mrs. Aara Will
John Morgan's girl baby

Roberts & Sons at 4 p.m. had
handled 120 dead bodies

29 Dead, in Carbondale Morgue

DEAD IN DE SOTO:
Luther Austin
Harold Hughes
Miss Hufhes
Joe Bullar
Margaret Neal
Joe Brothers Twins
Mrs. S.O. South
Mrs. Arthur Polson
William Schonkle
Bane Bridges
Ralph Dixon
Son of Harry Howell
Howell boy and girl
Lina M. Barnett

Frank Woods
Mrs. Fay Hide
Edna Estey
Fanie Sill
son of Jas Brown
daughter of Robt Brown
Leta Bowlaud
Mary Butterbush
Unidentified woman 45 yr.
Two boys
Frank Johnson
unidentified negro
Mr. & Mrs. J. Redd burned to death.

Many others taken to Du Quoin

An updated version of Laura Marie Miller's copy of Murphysboro's death list published by the local paper the day after the storm.

anybody whose house had been damaged or torn down. They did give some folks money—but we didn't need the Red Cross."

However, even a relief agency such as the Red Cross couldn't do anything to assist what was going on in the minds of the children of the storm, neither immediately after the storm nor in the decades following.

"I was frightened of later storms and am still frightened," Laura Marie Miller admitted. "There was a little old cherry tree outside my window and I was scared to sleep because I was afraid it would fall in on me. My parents got me over it by laughing at me and saying, 'That little tree isn't going to hurt you!'"

Laura Marie, however, three-quarters of a century after the event, didn't sound completely convinced.

Perhaps the extent of the disaster, and the long-term effects it left with its victims, could be summed up in the words of Mary Belle Melvin when she said, "Some things you think, 'Did I actually remember that, or did someone tell me? What all do you remember?'

"We didn't have any counseling nor anyone to help us through it. We were sent to my grandmother's house for awhile, and I was just eight, but I was put in charge of my squalling baby sister, who was two. *I* didn't want to take care of a crying little kid; *I* was a little kid myself; *I* wanted someone to take care of *me*."

But Mary Belle indicated that her grandmother had her hands full, so comfort was something that was at a premium in the days and months following the storm. It was only years later that many of the survivors and victims came to terms with the terrifying nightmare from the sky. And many, there should be no doubt, though they may not show it, never did come to terms with what happened to them or to Murphysboro.

DeSoto, Illinois
Dead: 69
Injured: 105
Percent of population dead and injured: 19
Property losses, 1925 dollars: $400,000

Perhaps one of the most distressing yet poignant tales of the Tri-State Tornado occurred in the little town of DeSoto, where the devastation happened on a large scale to the children. More than half the lives claimed by the insidious storm were children in the DeSoto schoolhouse—37 total.

Those children who survived were thrust into situations intolerable for even war-hardened veterans, let alone tender children whose limited life experience included school, home, families and love.

Lillian "Lala" Bridges Bullar knew nothing else but the staples of the American family until the storm came and mercilessly swept so much of it away.

Besides the severe injuries to her hands and fingers, in that they were burned so bad the flesh hung from them in strips, aesthetically, she had suffered greatly as well.

"My hair was burned and in such a mess," she said. "It was so long, all down my back, and had rubbish in it and they wanted to cut it off. But I wouldn't let them."

Lala's wounds were treated at the same hospital as her little brother Carl's—and, in the same facility where her mother died of injuries sustained when she was pulled by the raging winds from Benny and Edna South's home. Lala's father was devastated. His wife was gone, his children injured, some critically, and his home was scattered about DeSoto, piles of unidentifiable rubble.

In the face of all that, members of his family were approached by benefactor Harry Lancaster. Henry Bullar had found a guardian angel in Lancaster, who cared for the children as if they were his very own. Lancaster was the gentleman who had been kind enough to give Andrew Bullar and baby sister Ruby a ride to his fine home in DuQuoin—and to call his own personal nurse in to take care of them.

In the schoolhouse collapse, "Andrew had stuff blown all into both his ears," Lala said. "It was so packed in there, they waited a few days until it started to work itself out, but two or three weeks later, he was going crazy with it being in there and they finally washed his ears out three or four times, and got most of it."

Lala paused before she said, "We thought he hadn't been hurt. But that was bad enough."

Little brother Carl was in bad shape. He had places on his head where pieces of skull bone had been nicked out by the falling bricks of the school. As well, one eye had been pushed completely from its socket and was dangling on his cheek by connective tissue and nerves. But thanks to the magnanimous Mr. Lancaster, Carl was rushed in to surgery and the eye was put back in place, and repairs (exactly what Lala didn't know) were made to his skull.

She said she would never forget the day Carl was well enough to sit up in his hospital bed and regard the remaining members of the family.

"He sat up and looked around," Lala said tearfully, "and we were all there. He looked at me and Andrew, and Lola and Dad, and then he said, 'Where's Momma?'"

Lala couldn't repeat what his family told him; she didn't need to. Her tears said it all.

"He just lay back on the bed," she said.

The resignation the little boy felt was enough to make anyone understand.

"So Carl stayed there in the DuQuoin Hospital for three months," Lala said. "They fixed Dad the nicest big padded rocker and Carl wanted Dad to hold him and rock him."

Lancaster, said Lala, "done everything for us. He was a major stockholder in DuQuoin, practically owned the hospital, so what he said went. He took my dad in front of the staff there and said, 'Don't lock this door (to the hospital), day nor night. He has free coming or going day or night.'"

A view of the wind-destroyed De Soto School before cleanup commenced in the days following the March 18, 1925, storm.

Harry Lancaster even went the extra mile when he realized the Bullar family, were their father to go back to work anytime soon, was going to be children raising children, and he wished to lessen the load as much as he could.

"He tried to buy baby Ruby from Dad," Lala said with not a little wonder. "And Dad told him, 'Mr. Lancaster, all the money in the world couldn't buy her. I'd rather you'd have her more than anybody.' But…" Lala's words trailed at the thought.

She was proud of her father; that much was true. He wanted to keep the family together, and he succeeded.

There were other benefactors for the Bullars as well.

"We always went to church," said Lala. "We had lived east of Carbondale before we lived in DeSoto, and we went to the Pentecostal Church there. Soon as we got up there in DuQuoin they had the largest Pentecostal Church and they hunted us up.

"And the mine president came to Dad, and said 'Mr. Bullar, where do you work now?' Dad told him, 'I don't have a job right now.' The president said, 'Yes you do. I'm superintendent of the mines here outside DuQuoin' and told Dad to stay with the kids 'til we settled down and he got ready to start. And he

An unidentified man standson what's left of the foundation to his home in De Soto.

raised us kids, and me and Andrew stayed home with the little ones."

There seemed to be setbacks that no one could have foreseen; they weren't serious, but they were notable all the same.

"Ruby, God love her, got all mixed up," said Lala. "See, my name was hard for her to say. She'd say about me, 'That's my Auda' and that turned into Lala; and Lola was 'That's my momma.' From that time on, everybody in town knew me as Lala."

The tragedy was not only that the Bullars lost their house, but a couple of them, Lala being one, lost their sense of home in the village of DeSoto.

"I didn't want to come back to DeSoto and see that blood out there in the street. I knew it wouldn't still be there, but I wasn't sure I wasn't going to see it again. We did come back, but we didn't stay long.

"We moved to north of Dowell, and stayed there a year, then lived in DuQuoin before we built a house and went back to DeSoto," Lala said.

She went on with her life, eventually having friends and a boyfriend who later became her husband, but the tragedy remained etched deeply in her heart.

"Boy, I wanted my mother," Lala said so sorrowfully, even after many, many years.

The people of DeSoto pulled together like no other community.

Where houses stood, owners opened their doors to homeless residents in

A couple identified in local De Soto records only as "Mr. and Mrs. Wolff" sit amidst the ruins of their home in that town in the days following the storm.

order to give them a place to stay until help could come and relieve the agony of enduring another day without food, clothing, adequate shelter. Village president George Albon's home, a two-story brick structure, still stood, and Jeanette Ragsdale commented on the fact that Albon, her uncle, let many a homeless person board there until such time as the relief agencies could come and get shelters started. Tents arose all around DeSoto, but the rebuilding effort went on in earnest.

DeSotoans were resourceful. Caravans of cars that came from all directions were photographed in long, long lines coming into the village. But what the photographs couldn't show were what DeSotoans were doing with the drivers and occupants of the cars as they entered into the city limits.

It has been said, "The National Guard cracked down on Murphysboro, keeping the damage in and the help out." But further, DeSotoans claim "There was no National Guard in DeSoto. And the people of DeSoto welcomed any who arrived in their car to come see the damage, as long as they brought something with them into the town that could help anyone here in need. Clothing, food, personal possessions; if they had it, they were welcome, and if they didn't have it, they were welcome to come and help with the clean up and rebuilding.

De Soto being rebuilt thanks in part to the resourcefulness of its residents who asked those who would come and sight-see to bring and donate something to help the survivors – food, clothing, even money.

"DeSoto turned no one away."

West Frankfort, Illinois
Dead: 148
Injured: 410
Percent of population dead and injured: 3
Property losses, 1925 dollars: $800,000

Of the villages and towns experiencing death on a large scale, West Frankfort was second only to Murphysboro; 148 dead in the wake of the killer storm. It also experienced the second highest injury toll by number, at 410, behind Murphysboro's 623. Given such totals, it comes as somewhat of a surprise that the number of dead or injured in this major town was a mere three percent of the population, one of the smallest percentages on the list, but a statistic which shows the size of the town and the importance of it to the southern Illinois region.

For Arthur Reed, surviving the storm meant more than just living another day. He and his family had gone through utter chaos, they being unable to locate him for days following the disaster, and, having lost his baby sister Peggy Ann in a drowning which occurred during the storm, they were frantic as to where he could have been. However, he was discovered three days later, recovering in the Masonic Hall downtown—doing better than many of those around him, despite his having suffered a skull fracture.

His town, once a hub of industry, rail and progress, was in shambles in both residential and commercial areas.

A forlorn view of a West Frankfort neighborhood torn asunder by the Tri-State Tornado, with a baby carriage in the foreground.

"A neighbor had a house, but it was twisted around," Arthur said, indicating what an oddity it was for the house to be intact but swiveled on its foundation as if a giant child had come to it, picked it up like a toy, and, deciding against it, carelessly tossed the house back down where it fell.

"People had me stand in front of the damaged house and they took my picture. They were news people," Arthur said, delight obvious in his voice. "It was several days after I'd gotten out of the Masonic Hall, and they gave me fifty cents!"

The Red Cross came to be active in West Frankfort in the days, weeks and months following the storm.

More West Frankfort destruction, this of a neighborhood church.

"They brought in two tents for us, Army blankets, table and chairs and a cabinet," Arthur said of the relief agency. "Afterwards, we rented a home on West Poplar and the Red Cross billed Dad for all that stuff.

"Even in the service," Arthur continued, "in France during World War Two, they had 'donut dugouts.' The Salvation Army served 'em free. The Red Cross charged us."

*　　　*　　　*

The Red Cross came to the aid of the Gossett family as well, since their home had been picked up by the three-gust phenomenon and the foundation had sustained much damage on the final fall.

"The Red Cross bought Mom an oil stove," George said, but that was all the agency had to offer regarding their aid.

His father, beleaguered with the daunting task of repair to even the very foundation of their home, was hard-pressed to find temporary shelter in the aftermath. George did not make reference to any tents made available to the Gossett family. Instead, he brought out something a bit more unusual in regards to living arrangements.

"Dad found a house," George began, "and nobody was living in it, and we couldn't find the owner. So, we moved into it."

The family of seven needed shelter, albeit temporary, and no doubt Harry Gossett felt this was the refuge for his family until funds could be secured to reconstruct the foundation and repair other damage done to their own house.

The living arrangement worked until the owner discovered the squatters on his property many days after the storm.

The Red Cross supplies these types of tents for survivors of the tornado. Whole families lived in such structures throughout the year, keeping even their small farm animals outside, but nearby, such as the black checken perched on the ruins.

"He tried to vacate us, and the cops said 'leave 'em alone!'" George said. "But we didn't stay there long. We moved out a little later to another house on Taft Street, 'til our house was fixed."

The repair of their home was remarkable in its simplicity.

"They put the house back together, jacked it up, replaced the foundation and straightened the whole thing," George said.

George mentioned others who suffered the tornado's wrath that day—and not all of them were human.

"We had two terrier pups about seven months old who were missing a few days and came back uninjured," George revealed, in a rather amazed tone even after so many years. "Both lived about two more weeks and were killed by cars when we moved to the first house. We found the first one dead on the curb and I saw a bakery truck kill the second one a couple of days later. If I saw it today," George said, "I would say the driver killed the pup on purpose. He was in plain sight in the road. The driver had to have seen him but kept gaining speed and never stopped nor tried to."

But the worst casualties were the ones sustained by the 148 killed or the 410 injured in the town of West Frankfort.

"I remember a Mr. Moore who suffered a broken leg and afterward walked with a big limp. I also later knew Barney Clemons; he was later the County Clerk but he was a carpenter then, building on a house north of us. He

The strength of the winds that demolished the Gosset's home was able to roll entire flat cars from off the nearby rail line in West Frankfort. Below, a damage shot of a West Frankfort neighborhood showing a contemporary vehicle parked alongside the sidewalk.

was crippled and walked with crutches the rest of his life," George said.

"A friend Wilburn Bell and his sister Marie were found in their dead mother's arms near the Orient No. 2 Mine. Wilburn was my playmate when we were kids. He carried a big crooked scar on his forehead from the storm. He was killed along with 118 others in the Orient No. 2 coal mine explosion that occurred December 21, 1951.

"Marie, his sister, had to have a silver plate in her head because of how she was hurt in the storm, and died maybe fifty years later after having severe headaches for years."

After relating the thoughts about the death and destruction brought to

285

West Frankfort on March 18, 1925, George Gossett paused to laugh and remember one more particularly amusing, and quite amazing, tale he felt compelled to relate.

"My father's 1922 Ford was blown into the Grape Harbor area north of us," George remembered. "The garage was gone. And the car had very little damage; not even a flat tire."

Parrish, Illinois
Dead: 22
Injured: 60
Percent of population dead and injured: 30
Property losses, 1925 dollars: $77,000

Parrish, as residents in the early 1920s knew it, was gone following the winds and fires of March 18, 1925. It was never to return again as a viable community. After many decades, it didn't even have its own post office.

Adrian Dillon, only twelve when the monster storm passed through his life, was not naïve as to the repercussions the devastation brought to the once-thriving town. In his recollection of the more gruesome tasks that faced the survivors of the ordeal, his honest words reflected the impact the tornado had on the village and its victims.

"One of the most amazing things I remember about after that storm," he began, "was when it was over and thousands of chickens and cattle and dogs lay dead everywhere. Carcasses just *everywhere.*

"I remember there being swarms of men from Indiana coming in, and walking across the countryside, carrying shovels, digging holes for the carcasses and then wagons pulling lime coming to cover up and fill in the holes.

"It was a very important part of that deal," Adrian said simply, "because something had to be done with the dead bodies of those animals, but I didn't realize it as a kid.

"I just didn't realize it as a kid."

Adrian then drove home the point of the story. "As I got older, I thought lots about it."

It was obvious that as he grew up to understand the importance of the men from Indiana being there as part of the relief effort, there were things regarding *other* dead bodies, those not of animals, that were of equal importance as well. Animal carcasses were disposed of easily, even on such a broad scale as the one described. Human bodies, on the other hand, required more than what he next related happened to the animals.

"They dug holes, big as this room," he said and waved his hand around to indicate the vast expanse of the room in which he sat. The room used to be a large bank lobby, so the size was formidable, "and filled 'em up with the carcasses, sprinkled lime on it all, and went to another farm to do the same thing."

As for the human casualties, those injured or killed in the murderous winds or engulfed in flaming buildings, Adrian had one memory that stood out above all others:

"The rails were covered with trash, and it was hard going, but they had to get the trains in there, into Parrish. So they backed up a train from Thompsonville and Logan, and took out injured people."

He did not comment on what happened to the bodies of the dead; all he knew, as a twelve-year-old, was that trains were rolling out of his hometown carrying on them cargo that was once friends, neighbors, relatives, of all those he knew. Newspaper accounts from the days following the storm indicated the dead were shipped by rail, auto or wagon team to various morgues in the area, including those in Benton, West Frankfort, Thompsonville and other outlying towns with facilities capable of handling embalming and interment quickly.

Adrian's immediate family did not lose a member, nor were any of them seriously hurt in the storm. Though their house was gone, they were a farm family with enough wherewithal to be able to get back on their feet quickly.

"There were so many of us," Adrian acknowledged, "that we were scattered all over from staying with different people after that. It took two to three months to get everyone back together. I lived with an uncle close to Logan, Mom and the two younger kids lived with another relative nearby.

"We built back what we called 'the little house,' with one big room and two little bedrooms."

Adrian, however, did not get to enjoy being with his family in the little house for long. While he was staying with his uncle, he got the 'seven-year-itch,' (another term, and actually an 'old wives' tale' name, for eczema) and in order to avoid giving it to anyone else in such close quarters, he stayed out in the chicken coop, "nearly all winter," he said. Even though later it was discovered that the condition was not contagious, Adrian bucked up and did what he felt, at age 12, he had to do.

"I didn't want to give it to the other kids," Adrian said of his self-sacrificing move. "It didn't hurt my feelings. I didn't want anyone else to get it."

However, those with injuries that were a consequence of the storm did not fare so well in the months following as a boy with eczema did.

"There was a woman there in Parrish," Adrian explained, "who was sick in bed when that storm hit. And after that tornado blowed the stove over on her and nearly burned her leg off, she wouldn't stay in bed after that. She'd sleep someplace else, but not in a bed.

"A lot of people burned up in that storm. There were the Davis girls, they were daughters of Randall Davis, and oh, they were so pretty, two pretty little girls. One had nearly half her face burned off in the storm.

"Death rode the sky, alright," Adrian summarized of that day, in a sad tone of resignation that was repeated in the voice of one survivor after another. "I can close my eyes right now and see that tornado picking up the old Ross place, and just blowing it up."

Hamilton/White counties, Illinois
Dead: 65
Injured: 140
Percent of population dead and injured: unavailable
Property losses, 1925 dollars: $1,100,000

It was frequently said that what was remarkable about the damage done to Hamilton County was there were no concentrated population centers struck by the tornado, yet many deaths occurred in the farm population spread out in little communities all over the county.

With McLeansboro being the only large population center in the county, it naturally befell the survivors of the storm to bring their dead and injured to that little town. According to newspaper accounts and locals, the library, on the square in McLeansboro, was transformed into a Red Cross station, hospital, and temporary holding facility for those who lost homes to the storm.

The storm took the lives of 65 people, most of them members of farm families, or schoolchildren in the country schoolhouses that collapsed, like those in DeSoto before them, in the savage, tornadic winds. Farm families were particularly hard-hit, as to lose everything on their land and have their property ransacked by furious winds meant recovery was going to be long and bitter—if there were to be any recovery at all. So much debris was in the fields, said eyewitnesses, that spring planting was delayed and loss of revenue was the result.

One family who suffered more than any other ended up being that of the mysterious William S. and Rosetta Hicks Hollister, buried in the Albion Cumberland Presbyterian Cemetery between Enfield and Norris City, just across the county line from Hamilton into White. During 1999 production work on this book, newspaper press releases appeared in a White County paper, the *Carmi Times,* prompting response from those who knew the story of the relatively anonymous Hollisters to reveal the nature of their deaths in the winds of the storm, and the enigma of them in general.

"The Hollisters had moved to the Hamilton County area from far away; no one here really knew them, and they hadn't been here long. Rosetta Hicks Hollister was my grandma, and she and Mr. Hollister had been married only a few years at that time," explained Norma Brehm, a Carmi resident. "She had a daughter by her marriage to John Harrison Moore, and Rossetta was divorced from him before that baby was born. That was Marie Moore, who was my mother."

John Harrison Moore was Rosetta's second husband; Hollister., her third. Her first husband, Chris Hicks, had provided her with her first daughter, Amelia Hicks, who was called 'Medea' by all who knew her; Norma didn't know the reason why that name was what Amelia went by. Medea, 23, had married a man named Forrister, and was with her infant baby, sitting on the front porch of the home of her mother, Rossetta, 49, when the sky began to darken in the southwest on March 18, 1925.

Distant thunder rattled the chains that held the porch swing upon which Medea and her baby daughter sat. There with them on the porch was her mother, Rosstta Hicks Hollister. Medea's little sister Marie Moore, 9, had opted to stay home from school that day, but was not on the porch with the other two. Death's icy gaze watched them from the southwestern skies of Hamilton County. Marie had never been able to explain why, but, later in the day, she'd wanted to walk and meet the kids coming home from the nearby schoolhouse. Her mother, sister and baby niece sat on the porch and watched her go.

"The house was located near Dolan's Lake to the west of the Hamilton-White County line," Norma said.

The Hollisters were a farm family and were fairly prosperous. The property all around them was productive and ready for farming in the pre-spring days of March. Fields surrounded Marie Moore as she took off down the road. Apparently her family didn't feel much apprehension toward the approaching clouds; the nine-year-old seemed to know what she was doing.

The storm struck with its awesome savagery on the edge of Hamilton County at the Hollister's residence, reported Rossetta Hicks Hollister's granddaughter, Norma Brehm. It was so sudden and complete that the house was swept away, and everything on the property was almost fully cleared even of debris, according to accounts that had been passed down through the years to Norma.

"There was *nothing* left of that house," said Norma with emphasis. "They lost everything. And Grandma and Medea were killed immediately. The baby was blown away. I don't remember ever even seeing a marker for that baby. I don't know if they ever found her body.

"At first Rossetta was on the list of those just missing after the storm," Norma said. "No one knew what she died of. They found her body later on."

Marie Moore was located in a ditch bordering the property, with a severe cut in her head.

"It must have been pretty bad," said Norma. "Dad always said her head had been 'split open,' and she had hearing damage resulting from it.

"One of the Dolans was coming home in a wagon and found her," Norma said. "They put her in the wagon and took her to the doctor in McLeansboro."

With the Dolan's Lake area being on a more direct, westerly route straight into McLeansboro than many of the other areas stricken by the Tri-State Tornado, Mr. Dolan arrived with the little girl in tow sooner than anyone further south. Initially, he had taken her in her gravely wounded condition to the library in McLeansboro, in which the temporary care center had been located for victims of the storm. However, her condition was so serious workers there moved her to the hospital in town almost immediately. Therefore, Marie Moore carried the distinction with her through the years of being the first person from Hamilton County injured in the storm being admitted to the McLeansboro hospital, according to Norma.

*   *   *

Another enigmatic person killed in the storm about whom information later arrived, thanks to the *Carmi Times*, was the young man buried in the Enfield Cemetery, Vernon O. Miller, age 13 when he died.

A fine lady by the name of Rosey Miller called to inform that her late husband, Lelan Miller, grew up just south of Enfield with his family, which included his brother, Vernon.

"There at the crossroads of Illinois 14 and Highway 1, there was a schoolhouse called Trousdale School, and my husband and his brother were there," Rosey said.

The storm had just come across the Dolan's Lake area and was passing directly over the intersection of what would eventually become the two highways. After having devoured so many lives in Gorham, Murphysboro and DeSoto, it was as if the tornado were targeting country schoolhouses with their precious occupants, and White County would come to be no exception. The twister assaulted the schoolhouse, which sat to the north of the road just before approaching the crossroads from the east, before many knew what had happened.

"My husband's family lived west of the school, and there was another house nearby. That house is still standing," Rosey said of the fine two-story structure that was situated back off the road several yards.

"They had to dig Lelan out from under a lot of bricks; the chimney collapsed on him. He had a really bad cut above his ear on his head. He was nine when that happened, and he had a scar there forever.

"Vernon was the only one killed in that schoolhouse," Rosey explained quietly. "I didn't know either one of them then, I was only nine myself and lived in another part of the state, but Lelan told me Vernon had part of a two-by-four run into his back and through him."

Vernon Miller died at the scene of the devastation; the rest of the pupils survived their injuries.

Rosey remembered going to the Enfield Christian Church for the funerals that followed the carnage left in the wake of the storm.

"I remember seeing all these caskets lined up at the church. My mother's grandmother was killed, and she was there in one of those caskets."

The dead were conscious of nothing at all; they had gone to their final rest. As for the living, they were conscious of one thing in the aftermath of the trauma, if nothing else.

"All day, for days," Rosey said, "they were waiting on funerals."

Griffin, Indiana
Dead: 25
Injured: 202
Percent of population dead or injured: 60
Property losses, 1925 dollars: $375,000

There were two buildings left standing in Griffin at 4 p.m. on March 18, 1925. And 'standing,' as a descriptive term, was completely subjective under the circumstances.

"The Armstrong's house was damaged so badly, like it had been twisted a couple of times then set back down, but it was still upright. And the Murphy house was the other one standing, not hurt so bad, and a lot of injured people were being taken there," said Ellen Vanway Nottingham.

Her own family was being dispersed across the countryside. At first, Ellen and her surviving sister Evelyn were sent to the neighboring Brooks farm outside Griffin, where things weren't quite in the disarray most other farms were.

"We had no clothes," she recalled, "so the Brookses gave us long underwear and put us to bed. Grandpa Crawford came from Browns, Illinois, to bring us some clothes the next day. People gave them to him."

The outpouring of assistance from the neighboring community, while well intentioned, missed the mark for the little girl and several others like her, however.

"I remember things were too little or too big for me. There was a little yellow and white checked dress, and it was too small for me," she said.

One got the sense that Ellen wore it anyway, regardless of the fit; it was, after all, better than boys' clothes for the most part.

"But yellow was never my favorite color after that," she advised.

The Vanway children earned the distinction among Griffin residents as being the ones at whose house the school bus was stopped when the storm struck and carried many to their deaths. Ellen remembered more about the horrible day.

"Chick Oller, the bus driver, wasn't killed instantly," she revealed.

Apparently, according to rescuers, Oller was trapped inside the wreckage but worried more about his charges than of his own life-threatening injuries.

"He was telling the guys who helped, 'Get the kids first, get the kids!'" said Ellen, leading to the understanding that Oller, truly concerned for the well being of the children he was transporting, gave the ultimate sacrifice in the wake of the ultimate slaughter: He died while still pinned in the wreckage. So, however, did seven of the 30 or so children who were riding the bus that day. Ellen had little doubt that Oller's selflessness lead to the rescue and salvation of many of those more than twenty who survived.

However, seven children did perish, and little Ellen Vanway was not going to be permitted to forget the fact any time soon.

"You know, I told you the bus was stripped?" she questioned. "Well, they didn't do anything with it after the storm except pull it up out of the ditch to get Ethel Carl's and Helen Harris' bodies out from under the wheels."

The bus was then moved—to the Vanway's driveway.

"It sat there in our driveway for that summer," Ellen recalled. She didn't seem bitter or disturbed by a very obviously disturbing occurrence. "I had a picture of Evelyn and I sitting on the bus in the drive," she said. "I don't know what become of that…"

Her sister, Evelyn, who later came to be a resident of Browns, suffered permanently from her injuries sustained in the vicious winds and the debris within.

"Evelyn was on crutches for such a long time after that," Ellen said. "She almost died of gangrene from that wound on her knee."

It wasn't that the medical treatment was inadequate, but that it was slow in coming for the wounded and the dying in Griffin. Whatever facilities they had originally available to them, a doctor, a pharmacy, treatment-wise 'root' people who could have helped, were either dead, themselves injured or unable to attend to the severity of the injuries they beheld. Cut off from both ends, into and out of town, was the rail line that was the lifeblood of the community. Reports from the time indicate that when word got out of how severely Griffin had been stricken, relief trains were dispatched to the area from Olney and Centralia in Illinois and Evansville in Indiana—and the crews promptly found themselves unable to reach the stricken village due to the mess left on the tracks. Legend had it that complete train crews would get out ahead of the trains and clear the tracks by walking ahead. The train would inch along in a painfully slow crawl. There must have been a sense of urgency within the rescue trains and their crews and passengers of doctors and nurses, but the possibility existed that it was based only on curiosity. But it would have been interesting to learn what that curiosity turned to when the trains finally reached their destinations and discovered the ghastly remains of a town that had been a focal stop on the regular route of the Illinois Central rail line. The living were wandering with their pores soaked with river mud, blood seeping from wounds, a state of shock no doubt the norm. The dead were in no better shape than the muddied living; many of their bodies were covered with the silt and after a long enough time, the stench arose more quickly from those than others' bodies that weren't so stricken.

Ellen Vanway, as well as the other Griffin survivors interviewed who were mere children at the time, were unaware of how the train crews reacted in the face of such horror. All Ellen could tell of was the actual arrival of the relief effort by rail.

"Freight cars came into town later on, and they were where people could go if they needed something immediately," Ellen said. "There were some clothes to be had; I remember I got a little coat. And the Red Cross sent food in, but I don't remember it being a lot.

"The Red Cross did an awful lot for a lot of people," Ellen insisted, "but when it came to my family, they said, 'You've got good credit, go to the bank, get a loan.' They thought farmers could just go and do that.

"There was no help for farmers," Ellen said. "Not from the Red Cross. Now, the Farm Bureau, they were a big help. Our family wouldn't have had anything in the way of help for farming if it hadn't been for them. But with the Red Cross, it was a case by case thing." Ellen concluded by cautioning, "I have no criticism of the Red Cross. They did help a lot. Just not with the farm families."

Severe storms seemed to plague Ellen Vanway that particular year.

"I always got uneasy with storms after that," she admitted, "for so long we were so frightened."

Comparatively, 'uneasy' was a mild reaction to such a devastating occurrence; Ellen might have handled it better than others, but she had continuing reason to fear the storms as 1925 went on.

"There was such a bad electrical storm later on that summer," she said, "and during it, I took a terrible pain in my side at about 1 a.m. The next day at

6 a.m., I had an emergency appendectomy."

The weird weather of 1925 stood out in Ellen's mind for more reasons than just the nightmarish twister that screamed through her childhood home.

"It rained and hailed and lightninged something horrible after that storm came through," she said of the March 18 monster storm, "but the sun didn't set until just after 6 o'clock that night, and everything kind of cleared up and while it was sort of cold, there was one thing about it all, after everything."

What was that one thing?

"I don't know if it was because the storm came and swept everything clear in the sky, but that night was one of the most beautiful, colorful sunsets I've ever seen before or since."

There was no indication that Ellen perceived it as any kind of obscenity, this beauty following the massacre.

She just stated simply, "It was fantastic."

Princeton, Indiana
Dead: 45
Injured: 152
Percent of population dead and injured: 2
Property losses, 1925 dollars: $1,800,000

Princeton had seen far too many disasters in the early 1900s.

According to local historian and jeweler Greg Wright, there was a coal mine collapse in the early 1920s.

"There were men trapped underneath, in King's Mine," he said. "It was a full crew of twenty men, and a mule. The men got out by way of an airshaft. The mule didn't make it."

In late April 1925, about thirty days after the Tri-State Tornado twisted its deadly path across the countryside of Gibson County, an earthquake spawned from either the Vincennes or Wabash Valley Seismic Zone fault line rumbled the ground beneath Princeton and surrounding communities that the Wabash River bordered.

"It wasn't much," Greg said, "but it was enough to shake people. A man who had recently come to Princeton from outside the area was in town when the tornado hit. He was still there when the earthquake struck. After the earthquake, he decided he'd move away. It was all just too much to take."

From left, Dell Pride, Mrs. Eli Pittman and little Eloise Pride, survey the complete destruction that was the McClurkin farmhouse on March 19, 1925, one day after the tornado. The McClurkin girls' Uncle Art farmed the ground later, adn used to bring th em little toys he would find in the fields, blown there by the horrific winds during the March 18 storm.

The Tri-State Tornado, however, was the worst disaster to strike Princeton and indeed Gibson County. With what could be considered moderate death and injury tolls when compared to other major cities in the tornado's path, the numbers understated the impact on those who were left behind.

Little sisters Ada and Harriet McClurkin remained bedridden in Princeton's hospital for at least three weeks following the twister's attack. What they learned while they were hospitalized stayed with them in the decades that passed.

"We had a great uncle, a Doctor McClurkin, who was at a doctor's meeting in Evansville when he heard about the disaster. He hurried on and left the meeting," said Harriet.

"They wouldn't let him in when he came up to Princeton," she continued. "I guess he drove in, or tried to. So he went back to Evansville and came in on the first train he could get back. *Then* they let him in. I can't imagine why they turned a doctor away…"

"I imagine he said some choice words when they wouldn't let him in," Ada said.

Harriet looked at her sister and said, "Ada, he must have taken Mortin home with him, because he was gone for three weeks."

Mortin was the sisters' older brother. When asked if they missed him when he was gone with the doctor, both of them chimed in at the same time—"No!!"—and Ada felt compelled to add, "We had other things on our minds."

'Other things' included the loss of their mother, so soon after losing their father in a farm accident; the severe injuring of the aunt who eventually raised them, Kate Miller, who was hospitalized twice as long as the girls were; and the serious injuries the girls themselves sustained in the winds of the storm. These situations, in and of themselves, were enough to keep their minds off their older brother, whom they no doubt considered a pest, anyway.

And yet there *were* 'other things' to worry about.

"There was some looting," said Ada. "It seems like in a tragedy, people like that get what they want and don't care. You're down and they just kick you down farther."

Their homestead, or what was left of it, was a shambles. Barely a trace of the little farmhouse remained on the property. Still, said the sisters, there were items of household usefulness, pots and pans and ceramicware, which survived the winds intact and lay scattered about the property. Family members went to the farm in the days following, hoping to pick up whatever they could salvage for future use. And each time, there were less and less items to be had, gone, the sisters say, to the looters who came sifting through the debris hoping to find something useful for themselves.

"Uncle Art farmed the ground after," said Harriet. "Two different times he plowed up little play irons, like what I used to play house with, before the storm. He'd bring them back to me."

The thought of handing such an item to a little girl, who had lost her mother to the very winds which picked up her toy and flung it carelessly into an adjacent field only to be plowed up months later, might create a chill. But Harriet McClurkin was stoic. Or else, she looked at it with, simply, a little girl's perception of toys, and the joy they brought, regardless of their condition, or of the mighty killing force that put them where they were eventually found.

"I was glad he brought them back to me," she said honestly.

"But we never thought Mom would come back, though," she concluded. Harriet looked at Ada, who was nodding her head in agreement.

"It was okay. We really don't remember her."

*   *   *

A mother who survived was Arline Riley's.

Anna Whitten was looked upon by her daughter as nothing short of a heroine in the face of the loss and the tragedy the Tri-State Tornado brought upon their family.

She wasn't aware of the trauma her family had experienced prior to her birth until she was a little older, Arline said.

"I was born over two years after the Tri-State Tornado," Arline revealed. "Kids don't know nothin' from nothin.' I don't really remember a first time hearing the story. I just was aware of hearing about it, and that I lost a brother to it."

The Whittens made it a point to reconstruct their lives in every way possible, according to Arline.

"They built the house back where it had been," she recalled, "and I don't know if that affected Mom or not, but I know that she never got over storms, and because of that, I didn't like them either."

Still, Anna held her own and proved herself to be made of stronger stuff than the tornado could have taken from her.

"Mom was a very strong woman," Arline insisted. "When I was two, Dad got diabetes. He wasn't able to work for two years. So Mom, who had never worked outside the home, did laundry for twenty-five cents an hour.

"Oh, yes, she was strong," Arline concluded. "You know, she'd lost a baby girl at birth, and then Wallace, and lost everything of her home in the storm, so she had to be tough.

"She had to be strong."

Chapter Seventeen

The National Funeral Director's Association
"Every Man Did His Duty"

There are many aspects of a disaster on the magnitude of the Tri-State Tornado that to which, even upon careful consideration, the average human doesn't give much thought. The first response to such a disaster is generally the query of how many dead, how many injured, what is the extent of the damage in dollar amounts? Secondary is the rebuilding, the insurance and the assistance and the steps taken to finance the clean-up, and the effort involved in the restoration of normalcy and sanity when emerging from such an abnormal event and insane situation.

Those thought processes and logical progressions are the norm. But what of the details, the little things that no one thinks about until reality is staring them in the face? It is understood and has been established that several hours passed from the time the storm took out the farms, towns, and villages in some areas of the strike zone until rescue and help finally arrived in the form of doctors, relief teams, and supplies. Within those first hours, normal human processes, including those that happened among the injured, went on. People needed shelter from the elements that continued after the storm passed overhead. The elements included rain, hail, unconfirmed accounts of snow, and the inevitable temperature drop that set in at nightfall. Folks needed warmth that had been previously provided by coal burning stoves and fireplaces: in the best case scenario, those were rendered useless by the absence of a roof over their heads or the disappearance of chimneys and flues; in the worst case, the stoves had fallen through the floors of the dwellings and had burned the buildings and the people inside them, creating more hazards for the survivors; or, very simply, the stoves were something that the inconceivably high winds of the storm simply carried away.

Food was no doubt at a premium. In a day and age when simple meals didn't come in a can or a box, when refrigeration was a luxury, and many canned goods consisted of those that were prepared to the extent that they lost considerable nutritional value, food was an item that was beginning to take on a value of desperation within half a day of the killer storm. While many had relatives to whom they could turn in areas outside the strike zone, still more were completely destitute. How did they meet the basic needs for sustenance in this time of crisis? Were such food stores simply carried away in the ravaging winds of the twister? Did the survivors forage the wreckage of their own homes, looking for the familiar, only to find food stores in as much a state of destruction as the rest of their belongings? Or did they resort to their own brand of looting, a crime for which many were shot when doing so for seemingly malicious intent, as was evidenced in the Murphysboro account? Did they sift through the rubble of splintered homes and singed businesses, hoping to find bread or an unbroken can of vegetables or something perhaps more unsavory? How desperate did they get?

Simple hygiene took a back seat during the crisis. So many were covered with filth, soot, splinters, and in the case of those in Griffin, slime from the riverbed of the Wabash, that many were unrecognizable. Because of the grime that covered the youngsters sometimes head to foot, families overlooked their own children when they came to claim them amongst throngs of others. There were a few hand pumps that were capable of providing water to the masses of residents who no longer had a private place within their homes to clean themselves. There was no note made of toilet facilities or the lack thereof; that item of detail, no matter how uncomfortable it makes one only hearing of the tales, was imperative to the health and ultimate survival of many of the victims of the storm. Upon this reasoning, it becomes clear that hygiene was something that was not a minor concern in the face of the monumental disaster that had taken away so many and so much from those who survived.

And working through all these enormous problems, and on to the next logical progression of events over the course of days following the disaster, remained one final and ultimate question—

What did the survivors, many of whom had no facilities whatsoever to be used to deal with such a thing, do with their dead?

Bodies, as in the cases of Gorham and DeSoto, were literally lying in the streets following the devastation. Certain vast stretches of land, such as Cooksey Field in Parrish, became temporary outdoor morgues in which to lay out the dead to be identified by the living.

Others were not so fortunate as to have anything to identify. In Murphysboro, death reigned supreme in the northwest part of town through smoke, flame and fire. Within a Murphysboro motel, the charred bodies of those who took refuge in the basement were understandably difficult to identify. And within Murphysboro proper, as well as in other villages and cities in the other two states affected, there was always the issue of dismemberment. Body parts, unless they otherwise had scars, tattoos, or other types of identifying marks or jewelry/clothing, were generally useless in the identification process. As well, there was always a moroseness, on the part of the living, surrounding the need to identify the dead when all that remained was an arm, leg, or in some cases, a forlorn and macabre torso. Some simply did not have the intestinal fortitude to approach a severed body part and examine it closely enough to claim it belonged to a person known to them.

Within a matter of minutes, decomposition begins in a dead body, and within a matter of hours, the threat of disease to the living from the decomposing flesh becomes a very real one. In the case of farm families stricken by the tornado, they may have not only been at a great distance from the nearest undertaker, but left without reliable transportation or roads that were passable in order to get there. There also always existed the possibility that those who remained behind were injured, many severely, and were having a difficult enough time getting aid for themselves. So in some instances, the probability of disease from a decomposing corpse was particularly high.

In other instances, bodies were carried by a column of rotating air towering so high into the upper atmosphere that strong upper-level winds took hold of them and carried them wherever those prevailing upper level winds were blowing. They may have been suspended this way and traveled as a result for several miles, until they were deposited in an obscure location. Upon discovery and due to the passage of time, little was left behind of the remains. Some, in fact, were never discovered at all, being left to exposure to the elements in fields, forests, or other terrain. Several witnesses remarked that farmers, upon plowing fields in April and May during planting season 1925, came upon what was left of a decomposing body, or, in some cases, a mere skeleton.

Therefore, the question remains, what was done, given the scale of the disaster and the great toll it took upon the life of those in the three states affected?

It is a little known fact, these many decades after the disaster, that a certain group put forth an effort that required a considerable amount of determination, stamina, and compassion rarely beheld in mankind. The group was the National Funeral Directors Association.

A remarkable endeavor by area undertakers and funeral directors was launched in the hours and days following the catastrophic storm. The NFDA united their efforts from Missouri to Indiana and beyond, and prepared the way for the "embalming profession's fight against the spread of plague and pestilence in the devastated area…" While not all persons killed by the Tri-State Tornado were buried free of charge due to the NFDA, several in fact were, as their surviving family members were in most cases left destitute after the storm yet unable to avoid any of the preparations or trappings of a normal funeral and burial. The NFDA, along with the Red Cross, stepped in at this point to donate their services and resources free of charge, sometimes providing actual financial support to the victims' families; always going the extra mile to ensure that, during that time of incredible human crisis, the assurance that those departed would be cared for in a humane and respectful manner remained the one hope the survivors could have.

The set of articles found in issues of the journal *The Casket and Sunnyside*, a bi-monthly magazine that proclaimed itself to be "the two oldest undertaking journals in the world combined," revealed exactly what these ones 'undertook' to serve those in the disaster area. Published by Simeon Wile in New York City, the magazine claimed to guarantee "the largest circulation, each issue, of any undertaking trade journal in the world," and was a dignified, impressive document twice a month devoted to the bettering of the sometimes-maligned world of the undertaker and funeral director, someone everyone would need eventually after their lifetime ended. The entries from the April 1 and April 15, 1925 issues show attention to detail following the disaster that typified the dignity with which each individual undertaker and funeral director displayed toward their profession, and, especially, toward the disaster which struck the very heart of the country to which they tended so respectfully.

Coverage began with an editorial in the April 1 edition called "Every Man Did His Duty." In it, the writer stated, "The world has been shocked by the terrible tornado which recently swept through the Middle West (as the Midwest was termed back then), devastating whole towns and taking fearful toll of life in five states. The death roll in the wake of the cyclone of March 18 is as yet incomplete, for each succeeding day brings to light fresh horrors, and news from the stricken area comes to the rest of the country in disconnected driblets. To date, however, it is estimated that 102 persons lost their lives in Illinois, 98 in Indiana, 12 in Missouri, 2 in Kentucky and 30 in Tennessee. Added to this is a list totaling nearly 3,000 injured in the five states affected.

"In times such as this the first duty of society is to the living, and the greatest news interest centers about the survivors, injured and uninjured; but the dead, also, must be cared for. News of the manner in which the dead have been tended and by whom is as yet not available, but we are assured that the funeral directing profession has not been remiss in attending to its duties. For instance, we have this telegraphic assurance from President Harry M. Allen, of the Indiana Funeral Directors' association:

"'Our members giving all assistance possible to funeral directors in stricken area. Situation well in hand. Public's interest first thought of our members. Indiana association will also render financial aid.'

"Harry M. Kilpatrick, secretary of the Illinois State association, and of the National body, sends us this wire:

"'Radio station KYW, Chicago, at my request broadcast twice during Thursday, March 19, the following: "H.M. Kilpatrick, secretary, Illinois Funeral Directors' and Embalmers' Association requests embalmers to render all assistance possible in the area of yesterday's cyclone. Report at Murphysboro, West Frankfort, Benton and other places where dead are located. Listeners in are requested to notify their local embalmers." Responses received and embalmers proceeded to scene of devastation. Remained here, thinking I could do better service than by being in area, as all wire communication is yet out of commission. After two days' effort have just talked with President Zelle, of Murphysboro, who says embalmers have responded nobly and that they have plenty of help and are all right.'

"Like the story of a great battle, the story of the embalming profession's fight against the spread of plague and pestilence in the devastated area is yet to be told. Action comes first, report afterward. The Casket and Sunnyside will give the profession the detailed account of the noble efforts of the volunteers who answered the various calls for help, but at present we have but meagre facts in hand.

"Of one thing, however, we are certain. The embalming profession responded to the call for help as we knew it would. It has established another record to which it may point with pride. Every man did his duty."

In a second article in that issue, "Latest From the Great Tornado

Disaster," the staff wrote about Kilpatrick's radio broadcast requesting embalmers and funeral directors respond, with funeral supplies, to the storm-stricken area.

"The response was instantaneous and the funeral directors in the devastated area were soon surrounded by capable assistance and additional supplies," the editorial staff wrote.

"The storm occurred at about 2:45 p.m. Wednesday, March 18. Ambulances and funeral cars belonging to the funeral directors were first used to convey the sick and wounded to first-aid stations and hospitals. Later came the conveying of the dead to funeral homes and temporary morgues. That night bodies were embalmed under most trying and difficult circumstances—the water supply was destroyed, electric lights were out, with the result that torches, candles, lanterns and lamps had to be used. Caskets and embalming fluids and instruments were rushed in on the special relief trains, and by morning the work of caring for the dead was well under way."

On Friday, March 20, Executive Secretary of the NFDA Joseph N. Sletten, left Chicago for Murphysboro. He arrived that afternoon, volunteering his services to "the local funeral directors, the Citizens' Relief Committee, and to the Red Cross officials. Mr. Sletten was detailed to set up a system of graves registration and to supervise the digging of graves in the two adjoining cemeteries in Murphysboro. By nine o'clock the following morning approximately eighty men were digging graves for the 200 dead; squads of militia were acting as traffic directors in the cemeteries; and about 100 men with about 10 trucks were clearing the debris from the roads to the cemeteries in order that funeral cars might not be impeded. Ex-service men and laborers from nearby towns were imported for these duties. Ministers from nearby communities were also sent for to assist in conducting services in the cemeteries, as the Murphysboro clergy were busy ministering to the sick, wounded and homeless and doing other relief work.

"Except in extremely burned cases, every body was embalmed. Every body was interred in a separate grave. Every grave was provided with a temporary marker. Flowers and floral wreaths in profusion were sent by florists' associations in cities and by newspapers. In cases of extreme poverty, caskets were supplied by the Red Cross. In short, every body was properly buried.

"A.G. Zelle, President of the Illinois Funeral Directors' and Embalmers'

Association, is one of the three funeral directors in Murphysboro.

"This tornado, which in the twinkling of an eye killed over 800 people, injured over 2,000 others, and destroyed at least two million dollars' worth of property, teaches that no community can live—or die—unto itself. The spirit of the brotherhood of man was nobly demonstrated, and the funeral directors in the devastated area are grateful to all funeral directors, embalmers, funeral supply salesmen, drivers of funeral cars, and others who came to the scene of the disaster and assisted either contributing services, equipment and supplies."

An article headlining the April 15 issue of the magazine entitled "How the Profession Helped the Storm Sufferers" continued to outline steps the NFDA took to ensure the dead along the tornado's path were given the dignified and respectful treatment shown to anyone under any different kind of fatal circumstance, a remarkable feat given the horrific conditions following the storm's furor.

In this particular article, a much more widespread view was given of the weather events of March 18, 1925, including attendant weather events in other states besides the Tri-State Tornado's three, events that spun their own tornadic systems and created death and destruction, albeit on a smaller scale, in states besides Missouri, Illinois and Indiana. This account is one of the few publications that viewed the nation's weather activity on that day as a single, far-reaching event, instead of isolated incidents that were relegated to the pages of history as they lay gathering dust in local newspaper archive storage rooms or photocopied accounts in libraries and historical societies.

"What was probably the worst calamity due to natural causes which ever visited the United States occurred on March 18 when a tornado of unprecedented fury and extent swept through five states killing nine hundred and ninety people, injuring nearly six thousand more and doing incalculable property damage," magazine staff wrote.

"According to the most accurate figures thus far obtainable the death score in the stricken communities was as follows: Illinois: West Frankfort, 102; Murphysboro, 150; De Soto, 158; Vergennes, 38; Gorham, 35; Parrish, 25; Royalton, 3; Carbondale, 25; McLeansville, 18; Carmi, 2; Blumfield, 7; Caldwell, 9; Logan, 10; Thompsonville, 9; Rural Districts (thus far accounted for), 50.

"Indiana: Princeton, 98; Owensville, 64; Poseyville, 5; Griffin, 65; Elizabeth, 3.

"Missouri: Cape Girardeau, 12; Altenburg, 1; Annapolis, 3; Biehle, 10.

"Kentucky; Springfield, 2; Lexington, 1; Buck Lodge, 3; Keytown, 2.

"Tennessee: Gallatin, 30; Angle, 15; Witham, 12; Oak Grove, 10; Sulphuria, 8; Buck Lodge; Peatown, 2.

"Total dead: 990."

The compilation of these locations and figures are fraught with mistakes, as are apparent in the spellings ("Blumfield" for Plumfield, "McLeansville" for McLeansboro) as well as locations (the tornado did not strike McLeansboro, as well as moved considerably north of Carbondale) but these may have been compiled by *residence* of those deceased in the storm; in other words, from which village or town they hailed, as opposed to where they were found when their deaths were counted.

The account went on to state, "To these appalling figures must be added a total of 5,783 persons known to be seriously injured. The death total will undoubtedly swell when those fatally injured, but still living, shall have passed from their sufferings.

"Great as was the disaster, the spirit of cooperation with which the organized funeral directing profession met its challenge was even greater."

The article goes on to recount the "story of the official activities" of Harry M. Kilpatrick, secretary-treasurer of the National Association, and secretary of the Illinois State Association.

"When I first learned of the terrible storm and cyclone in Southern Illinois I realized at once the necessity of prompt action in assembling a group of embalmers to assist the local funeral directors in caring for and burying the dead," Kilpatrick was quoted. "I knew, beyond a doubt, that the embalmers of the state would respond willingly and render every service possible. I realized, too, that many hearts would beat in unison in sympathy for those afflicted in the storm area, and also that all members would be intensely interested to know whether or not our beloved president, A.C. Zelle, of Murphysboro, was among the long list of dead.

"I immediately tried to get in wire contact with President Zelle, Murphysboro, Vice-President A.G. Storme, Herrin, and several others, including officers of the St. Louis association, but without success. I then put in a call for radio station KYW Chicago. It was not possible to get the call through until shortly after 12 o'clock Thursday, March 19th. At my request, KYW broadcast at 2 p.m., and also once later during the afternoon the following message:

"'H.M. Kilpatrick, secretary, Illinois Funeral Directors' and Embalmers' Association, requests all embalmers to render all assistance possible in the area of yesterday's cyclone. Report at Murphysboro, West Frankfort, Benton and other points where the dead are located. Listeners in area requested to notify local embalmers.'

"These broadcasts brought about a remarkable response," Kilpatrick was quoted. "Immediately telegrams and telephone calls came in the office when instructions were given asking each one to report at once, prepared for the service in view. Fully fifty or sixty funeral directors and embalmers rushed to the scene of the disaster, and rendered valiant service."

Kilpatrick told of contacting Sletten, and how Sletten set up the graves registration system among the citizens "and rendered splendid service.

"When I first learned of the great disaster, my desire was to immediately proceed to the stricken area," Kilpatrick was quoted, "but when I learned that all wire communication was cut off, I realized if I got in the storm zone I could not communicate with those outside, so decided that I could represent the Association, and render better service to remain at my desk and try to suggest and direct. It is not possible to tell the names of all who so cheerfully responded. I received calls from Arkansas, Wisconsin and many from Illinois. Every effort was made to get embalmers on the ground.

"I did not succeed in getting in direct contact with President Zelle until about noon Saturday, March 21. Then I was advised that they had all the assistance needed and that everyone was giving the best of cooperation.

"A wire report from West Frankfort gave the same information, further stating that all bodies there had been properly embalmed.

"All those who responded to the call and rendered service are to be commended for their action and are deserving of the greatest praise and thanks. Such loyal and valiant service will undoubtedly raise our profession in the minds of the public."

The article reiterated points made previously in the April 1 edition about how each body was embalmed and interred in separate graves, "properly buried."

Then the article gave a rare glimpse into what happened in states affected by the weather outbreak of the day—and not directly by the Tri-State Tornado itself. These attendant storm systems were no less horrific to those they afflicted than that which spawned the Tri-State Tornado, despite the lower numbers of dead and injured, and as it turned out, this account shows a much wider area was impacted by weather on that March day than Missouri, Illinois and Indiana:

"H.S. North, of the firm of Harris and North, Gallatin, Tennessee, tells this story:

"'The storm passed through the northeast part of the county, known as the Ridge section, which is a hilly section of country. The destruction started about eight miles north of Gallatin and moved in a northeastern direction across the county, covering a space about five hundred feet wide and about sixteen miles long, destroying houses and timber. There were about forty houses and barns destroyed and an untold number of head of stock and cattle and chickens scattered all over the ground close to every house that was destroyed'," the article related in what brings to mind the scene of devastation of unfortunate Parrish, Illinois.

"There are thirty-three known deaths and about seventy-five injured," the article recounted North saying. "About forty of the seventy-five injured are in hospitals and the balance are not considered seriously enough injured to send to hospitals. There was an emergency hospital opened at Gallatin by the local Red Cross (which has no hospital) to take care of the less serious cases, and the cases needing surgical attention were sent to Nashville, twenty-eight miles away.

"The property damage is estimated between one hundred and twenty-five to one hundred and fifty thousand dollars.

"There was no damage to any member of the undertaking profession in Gallatin. The profession was able to handle all cases in a professional way. Two ambulances running night and day, with touring cars for the less seriously injured, were able to handle the situation without much delay. The dead were handled in a professional way and each and every one was furnished a casket, not one buried without being placed in a casket, although three or four were so badly mutilated the bodies were not made presentable."

The article issued the names of those providing ambulances (Wilkerson Brothers and Wiseman, of Portland, Tenn., and Harris and North of Gallatin) and spoke of the WLS broadcast requesting help for those in the afflicted areas—even for far-flung Tennessee—and how wonderful the response was, with rail cars bringing supplies, food and clothing as assistance to those stricken.

"C.C. Alexander, of Gallatin, Tennessee, wrote: 'I had charge of ten bodies,'" the article continued. "'The storm passed through the northern part of Sumner County about half-past five o'clock, and I was called a few minutes before twelve, midnight, to come to the storm area. I went and worked until just before daylight, getting bodies from under timbers of wrecked houses and from the woods, and taking them to a home nearby which had not been blown away.

"'Four of the bodies which I handled had been torn apart at the hips, only the trunks being found.

"'I called my son, who has been in the profession at Lexington for six years, to come and help me, but he was unable to come until Friday following the storm. I buried the four badly mutilated bodies on Thursday afternoon and, with the aid of my son, buried the other six on Friday," Alexander recounted.

He went on to reiterate what the task of the NFDA members had been all along: to prepare for burial, provide services, and give each corpse a respective grave.

The article then mentioned that the Chicago Undertakers' Association wired Murphysboro, Ill., with their "profound sympathies" and noted "our services are at your disposal. We are cooperating with a committee appointed by Mayor Dever, members of which have pledged themselves to collect five hundred thousand dollars."

A brief mention was then made of the devastation in Indiana:

"From Porter F. Short, of Owensville, Indiana, came this terrible message: 'Owensville dead 24, injured 150. Farming community total loss. Injured in Evansville hospitals. Town under martial law. One hundred thousand sightseers through area on Sunday. Griffin dead 47 including Mr. and Mrs. Charles Stallings, local funeral director. Six missing. Town totally destroyed.'

"Harry M. Allen, of Peru, president of the Indiana Funeral Directors' Association, gave this report of indomitable spirit: 'Our members giving all assistance possible to funeral directors in stricken area. Situation well in hand. Public's interest first thought of our members. Indiana association will also render financial aid.'

"This is the story of the manner in which the American funeral directors, singly, as individuals, or co-operatively, as associations, rendered assistance to the stricken areas," the article wrapped up. "Later the full details of the stalwart work done by members of the profession can be told at greater length. Today the confusion surrounding the ruins of what were once happy homes is too great for anything but the barest outlines of the story to be known. One fact, however, stands out from all the surrounding confusion. Every man in the undertaking profession did his duty."

An article in the April 15 issue called "A Note of Preparedness," written by Captain Henry S. McCrea, of Worcester, Massachusetts read in the preface, "Steps to be taken by associations and individual funeral directors in case of other disasters similar to the tornado which devastated the Middle West last month.

"Without assuming the attitude of an alarmist I wish to commend the Casket and Sunnyside sounding a note of preparedness to the funeral director and sanitarians of this country in the event of public disasters such as that which visited a section of the Middle West a short time ago."

McCrea noted that where his offices were located along the Eastern seaboard, they sometimes give little thought to the magnitude of weather disaster the "Middle West" often faced with spring storms.

He then went on to speak of the challenges his agents faced regarding public perception, something that was very important to the NFDA, long before "emergency management" and the dramatization of disasters in order to prop up televised reports—and thus ratings—was but a fleeting thought. That sincerity of the profession stands in stark contrast to those who actively seek publicity in order to continue being funded, either by government means, or by means of celebrity.

"Individually or collectively we should arrange at once to meet the health authorities of our cities and let them know that our service and equipment is at their command free of charge in the hour of a public disaster," McCrea wrote. "This public spirited movement would disarm any cry of propaganda that would connect us with ambulance chasers and ghouls when we were called upon to handle charred and mangled bodies within the police lines.

"As a reserve officer in the Quartermaster Corps," McCrea continued, "I should feel in duty bound unofficially to assume the responsibility of using the cadre of Graves Registration Unit No. 349 in any emergency measure and as commander of that unit I would offer the services of my trained men to the public health and Red Cross officials as well as solicit the cooperation of every funeral director and embalmer within the zone of operation.

"It would be impossible to estimate the percentage of bodies that could be embalmed and disposed of according to regular funeral customs," McCrea referenced a disaster the size and scope of what lay in the wake of the Tri-State Tornado, "but with the approval of the medical examiner the work could proceed without delay by a detail of embalmers irrespective of the family choice of a funeral director later on, and for once at least we might find ourselves helping each other.

"In conclusion, I will add that it should be the duty of each state and municipal board of health to check up on the practical knowledge and equipment of every man in the profession to ascertain his capacity to cooperate with them in the event of a public disaster," McCrea wrote.

With no grandstanding, no expectation of receiving recompense, no demand for media coverage and no one watching but the survivors of the devastation, the National Funeral Directors' Association did what their hearts, devoted to the dignity of their profession, told them to do: handle the results of the disaster in the most expedient way possible, with the decorum that those in their field were accustomed to providing.

Their thoughtfulness and devotion was not lost on the populace who were aware of the solemnity that came with the occasion; even 75 years later, when information for this book was being compiled, several survivors mentioned the heartfelt gratitude their elders expressed toward the care given their deceased family members and friends, even in the harrowing days post-storm, when the magnitude of the dead, and the urgency to begin interment, pressed upon their sorrow.

Thanks to response from the National Funeral Directors' Association in Brookfield, Wisconsin, the articles from the *Casket and Sunnyside* publications from their archives carefully preserved at the NFDA's Howard C. Raether Library were provided upon inquiry for this book. These accounts can be added to those of the survivors of one of the greatest disasters ever to strike the United States…and they provide an eloquent testimony, in the words of those responsible for carrying out the saddest of all eventualities with compassion and care, of a day and time in which sometimes the only reward was the gratitude of those left to mourn.

Chapter Eighteen

The Weathermen

*"The fact is that these synoptic-scale cells are
infrequent, or we'd have more of these beasts"—
Dr. Stanley Changnon Jr., Author, Chief Emeritus
at Illinois Water Survey, business owner and former
Professor of Atmospheric Science and Geology,
University of Illinois, Champaign*

After having interviewed over 50 survivors or direct descendants of
survivors of the Tri-State Tornado of 1925, it was easy to fall into the same
pattern of thinking that they had adopted over the years regarding the storm.

One of the primary elements of the thinking was so much legend and lore
involved in relation to how victims of the nightmare viewed this and other
storms. For centuries, tales abounded in which there were certain locations of
terrain that were, by turns, either cursed or blessed according to how the
weather struck them. Native Americans, in their belief systems based upon a
Great Spirit and the spirits of objects both animate and inanimate all around
them, felt that tornadoes were a natural part of the cleansing ritual nature
provided for them. Some, however, also looked upon such storms as the Great
Spirit sweeping down upon them seeking revenge—or worse for the new
settlers in the land, a sweeping revenge upon *them* for being interlopers upon
Native—and thus sacred—ground.

Conversely, many such cultures, as previously noted, felt that certain areas
of the country were more protected than cursed by the forms of weather they
received. Particularly when tornadoes or other damaging storms would make
touchdown with every area outside a certain Native encampment except the
very area where the Natives resided, this would cause them to think that there
must be some kind of blessing upon the land on which they chose to live. It
didn't matter that over a period of decades the Natives may be more nomadic
than anything and selected different areas year after year, and potentially not

see the destruction that a storm may have brought to the spot they just left. They sometimes felt, perhaps justifiably so, that their very presence upon the land created a talisman that could thwart damaging storms.

In later American history, local legend played into the sentiments the residents had toward such violent weather. Even in the absence of the Natives when settlers took over Missouri, Illinois and Indiana, the locals had their own way of dealing with the way the weather itself dealt with them. In the Ozarks, it was noted time and again that storms in the hills and valleys were prone to "building," gaining speed and momentum in the rising and the falling of the landscape. Such storms, noted the locals, were known for their ferocity, and were dreaded in southeastern Missouri. These theories were presented in such a decisive and definitive manner that the uneducated would almost naturally assume the accuracy of the tales without so much as a doubt.

This was the case upon hearing the early tales of the Tri-State Tornado as it made its treacherous way across the first leg of the journey, the foothills of the Missouri Ozarks. The Foxes, Sam Flowers' nephew Jim, Clara Brown and Peachy Jones all displayed the thought process that led to assertions of storms taking hold of the hills and hollows and using them to build momentum and fury as they passed over. And to look upon the photos of the destruction of such a place as Annapolis, Mo., after the Tri-State Tornado swept through, might lend credence to that belief system.

However, an Illinois atmospheric scientist, Dr. Stanley Changnon, Jr., of Champaign, was of a different mindset.

"The Tri-State Tornado is in a family of unique long-track events that develop deep localized low-pressure systems and will create a cell that will be the mother of the tornado," he said. "It's not the Ozarks—it's the fact that these synoptic-scale cells are infrequent or we'd have more of these beasts."

Changnon acknowledged that legends and mythology run rampant where such giant storms and other weather phenomena are present.

"People in this country hundreds of years ago were totally victimized by the weather," he said. "They developed a huge amount of folklore to help them try to understand what was going on."

This, he indicated, was likely what lead to the foundations of long-held beliefs about such a storm as the Tri-State Tornado, and of why it had such deadly results.

Stanley Changnon, Jr., Ph.D.

Changnon, as the Professor of Atmospheric Science and Geology at the University of Illinois at Champaign/Urbana, had spent several years of his career studying weather and its impact on the state of Illinois. For the book *Illinois Tornadoes*, Changnon, along with John W. Wilson, made an in-depth study of the Tri-State Tornado that included reconstruction and charting of the path. This map, showing the southern portions of Missouri, Illinois and Indiana, details the track of the storm from its beginnings outside Ellington, Mo., what time the tornado made landfall with each village affected, and the forward moving speed from area to area, based upon the distance traveled and time of impact. Changnon reiterated that for months after the 1925 storm, there was a major survey done of the track by engineers.

"That unto itself was a miracle of doing the right thing—at a time when that studying of unusual weather wasn't done," he commented.

The help the survey provided future meteorologists and atmospheric scientists, along with countless other interested groups and individuals, was impressive to Changnon, who credits the engineers with aiding a present-day clearer understanding of the storm, and of its mighty impact upon that corridor of the nation.

"But we need to take a longer view or a different view of questions regarding how the development of the storm in the Ozarks happened," said Changnon.

He discounted the idea of the theory that storms 'build' within the dipping and rising of the hills that characterize the landscape of that part of Missouri, and indicated that there were so many more factors that affected the strength of the Tri-State Tornado that hills and hollows could not have impacted in any way, shape or form.

"Most very long-track tornadoes in the United States are generally east of the Mississippi River. Even the 'tornado alley' of the U.S., which is actually the Oklahoma, Kansas and Missouri area, has never had the long-track type of tornado we're talking about here. The critical part of this thunderstorm development west of the Mississippi is that this was something that had a long lifetime. Most storms last 35 to 75 minutes—but this one lasted hours.

"That factor required a localized but strong low pressure system that allowed development and encouraged the tornado to last a long time," Changnon said.

Hills, hollows, the rising or falling of the land—none of it would have made any difference to a storm of that size and magnitude.

"That funnel will keep going. Nothing would get in its way," Changnon concluded.

Ernie Kern, who in 1999 was Professor of Geological Sciences at Southeastern Missouri University in Cape Girardeau, voiced a similar opinion, basically debunking the 'building' theory of storms within the Ozarks landscape.

"The influence of topography on tornadoes is not well-documented," Kern said. "The general consensus is that topography plays a relatively minor role on the level of force.

"Now, you do hear a lot about a tornado getting caught in a valley," Kern admitted. However, he would not assign any level of importance to this theory over the 'building' theory. Instead, Kern was more emphatic on revealing what factors created the tornado effect to begin with.

"The controlling factor of a tornado is a thundercloud," he explained. "Large cumulonimbus clouds that may begin as far up as only 500 feet."

Thus, he pointed out, such low clouds hanging over valleys may *appear* as though they are following a natural land trough—but this isn't necessarily the case, either.

"Those cumulonimbus clouds may go up to 40,000 feet high. So the mechanics of tornado origin are all aloft. They typically form in squall lines in advance of very intense cold fronts, with very warm air clashing with very cold. Oftentimes, and really, usually, the warm air is unstable.

"The normal tornadoes have developed out of plain, isolated thunderclouds," said Kern, "and most tornadoes form in places in easy access of two air masses from the north and south—like a huge bowling alley."

Therefore, some low-lying land areas could conceivably appear as though they are fostering the development of storms directly above them, when all the observer is seeing is a more clear line of clouds due to the observer's position of them from the earth.

Was it possible for the residents of the Ozarks to have been unable to recognize the Tri-State Tornado for what it was, as indicated by eyewitnesses who spoke of only a 'wall of fog' roaring ahead as it smashed through the hills?

"That could be, depending upon their location as compared to the tornado," said Kern, "and when they first sighted it. The hills could have been blocking their view of the horizon, so it may have been right upon them before they could see it. Then add a lot of heavy rain and clouds and it would have looked like a fog and not an individual funnel."

The appearance of the Tri-State Tornado had long fascinated both eyewitnesses and those interested who were not present for the phenomenon. Reports of a 'truncated, inverted cone' repeated from east of Biehle, Missouri to just before Griffin, Indiana. Truncated, in this case, meant that there were levels of clouds in a descending pattern, that appeared as though they were chopped straight off. Then another level appeared, then another chop, and another level, and so on—rather like an upside-down version of the 'tiling' of hillsides in order to plant properly on the leveled earth. Thus, the Tri-State Tornado was highly unusual in effect and appearance.

Kern addressed the latter.

"Some tornadoes have no funnel, just an undefined appearance. Along the base of the tornado is the contact with the surface of the earth, and there you'll find a tremendous amount of debris flying, not being retained right at the tip, like an explosion throwing it around," Kern said. "I've seen several cases of tornadoes where the bottom looks like an upside-down mushroom."

Therefore, he noted, such an upside-down 'cone' was entirely feasible in the case of the Tri-State Tornado—the base, or debris field, could have been of such a magnitude that it actually appeared wider than the massive cloud formation at the top of the column from which the pillar was originating.

Appearance of the storm as it moved across the two rivers dividing the states was another point upon which eyewitnesses seemed to fixate. In Gorham, it was said the tornado grew to massive proportions as it crossed the Mississippi, drawing up water and everything within the river as it passed through. The same expressions were made in Griffin, as, having crossed the Wabash, the local legend had it that the riverbed was temporarily sucked dry at the point of impact as river water was drawn up with the tornado's force.

"Rivers are more folklore," said Changnon on the subject. "The dynamics of a tornado don't react on that kind of space and time scale—there's only an extensive deep low pressure center moving and sustaining itself. Bluffs, rivers, things like that, would make no difference in the impact or appearance that tornado would have had."

"It could *skip* across the top of the water," said Kern, "and it could pick up lots of water, and if you have a major tornado moving at a comparatively slow speed, it could remove water from the (river) area temporarily."

But to have a complete removal, down to the riverbed, would be virtually impossible, noted Kern, considering the volume and pressure in a constantly flowing body of water the size of the Mississippi River, or even the relatively smaller Wabash.

"The water can compact somewhat and it will leave a trough, but it probably won't get to the bottom of the river," Kern said.

Reports from old-timers in Grayville, Ill., at the Wabash, as an example, of being able to see the riverbed when the tornado passed through might not be complete fabrication, according to Kern, before the rushing of the water came split seconds later and filled the trough back in. But the likelihood of the 'river being sucked dry' as a result of the Tri-State Tornado, and staying that way for any notable period of time, was probably exaggeration at best.

And of the image of water being driven ahead of the storm as it crossed the Mississippi at Gorham: this might have been provided by something other than water being drawn up into the vortex, as many eyewitnesses presumed that it was, according to Kern.

"That might have been condensed water vapor," said Kern. "It's known as 'ancillary' (or secondary) condensation, which can happen all along the fringes of the storm."

The force of the winds within and surrounding the vortex, in other words, would not necessarily drive the water from its place in the river so much as change the state of the water as it churned, much like watching a blender whirl plain water into a froth at a high rate of speed and in a confined location inside the blender itself. The river water, implied Kern, was simply vaporizing within the vortex of the tornado.

Kern also addressed unusual aspects of the storm, such as a 'vertical horizon' or flat wall of clouds moving along the ground.

"Now, the 'clouds lying on the ground' theme is very possible, especially after a crossing of a major body of water," said Kern. "It goes back to the debris field and what explosions are going on at that point. It's sometimes hard to tell the difference between where the dust kicks up and where the clouds begin. The same thing, then, can happen with water as the vaporizing is taking place."

The menacing appearance, then, of the storm as it crossed the bodies of water, was a valid one.

What happened to the water, being driven forward and outward or being vaporized, was debatable. What happened to what was in or beneath the water, i.e., fish, muck and mud, however, was commonly known. Fish pelted the earth in places sometimes far from the river as they were flung into the heart of the storm and back out; muck was driven with merciless force into the pores of people, both casualties and survivors, in and around Griffin, Indiana.

Among the most fascinating points eyewitnesses maintained was the "three-gust" effect the wind had on them during the experience of the tornado passing overhead. This phenomenon was generally limited to those who were directly under the actual vortex itself. Sometimes it resulted in devastating consequences, as in the case of Lala Bullar Bridges in DeSoto. Sometimes it resulted in only a moderate but very discernible sensation, as in when Parrish resident Lena Ross commented that "the air got close three times."

Both Changnon and Kern approached this effect with caution. Changnon believed it to be more of an effect than a distinct phenomenon, as he said, "A rational explanation would be the most strong thunderstorms with or without a tornado can create an initial downdraft that can be very strong. Air pouring out (of this downdraft) gives you a sharp, close-to-the-storm feeling.

"Then, passage of a funnel, depending on which side you are on, can give you winds from different directions. You can be on the east side, then suddenly the west side of the winds of a funnel as it passes."

Changnon continued, "People were experiencing different, normal facets of a storm. They were feeling and seeing the initial storm appearance; then stillness; then with the passage of a funnel depending on which side of a mile-wide storm you were on, different winds of incredible pressure… then all of a sudden they would be on the other side of that mile-wide storm, and everything would feel different."

The phenomenon, said Changnon, would vary from place to place and person to person. Therefore, it was his opinion that the three-gust theory had its basis in the dynamics of such a monster storm, and not necessarily in the elements contained within the storm itself.

The latter explanation, however, was what Kern took to task when presenting his observation on what created the three-gust feeling, from Peachy Jones' perception of it in Annapolis, Mo., (describing it in three stages as she said "It was light and it was dark and it was over") to Hal Davenport in Crossville, Ill., telling how the "whole cellar raised up and down," as if the storm were trying to "pull them out of the ground," perhaps more than one time but no more than three.

"There's a theory that a tornado is more than just a single entity," said Kern. He used an illustration. "If you draw a circle, imagine you're looking straight down the funnel from the top; that's what you're seeing. Then within that, draw three to five smaller circles inside that circle. You're still looking down from the air above the funnel; within the funnel itself there are now other suction spots.

"These are rotating around within the main funnel, but actually, they are tremendously high winds within the suction spots," Kern said.

The entire rotation is changing at this point, he said, and it becomes hit or miss as to whether the suction spots will strike an object that will suffer devastation from being in its path. Or, it may depend on whether the rotation will be such that the suction spot will be away from a building or tree, and effectively "miss" it, even though the same building or tree is actually under the total rotation.

Kern explained it further: "This whole tornado could be seen, for example, passing over a grain store, a factory and a volunteer fire department all on the same block. The grain store is torn up, the factory is obliterated and the fire department hasn't got a scratch because by chance, one of the suction spots didn't encounter that building, but it did the factory, and just brushed the grain store, as it passed overhead."

In a storm the size of the Tri-State Tornado, there could have been more than three suction spots within a giant, rotating column, but with witness reports being what they were, three seemed to be the number of the day. Each time a witness described the air getting "close," or the "first gust of wind did this… the second did this… and the third did that," reasoning dictates that there were three elements within one, and that each element was making its own impact upon the ground surface as well as what it was picking up within the vortex and flinging around.

As for multiple tracks, these had been confirmed in the weeks following the storm by civil engineers who made a detailed study of the path, as previously indicated by Changnon. Path width could only be defined, said Changnon, by a damage track within a populated area immediately following the storm. As for wide-open spaces (Hamilton County's rural area, as an example) it was impossible to determine what was damaged directly by the tornado as opposed to what was peripheral damage caused by objects being flung into the area. When the swath was cut through a town or community with a more dense population, as in Murphysboro, it could be measured with relative accuracy and a single, double or triple path could be ascertained at that point if eyewitnesses hadn't actually observed multiple funnels.

Changnon addressed the fact that in some cases no funnel was observed at all.

"The monster was hidden, given the lack of experience with tornadic storms that the people in the path had," he said. "They didn't know what they were looking at. It didn't look like a tornado."

From reports in West Frankfort that Changnon took personally over 40 years later, he said, "Folks at Orient No. 2 said it looked like a big black wall, like dirt…nobody at that stage saw a funnel."

Other measurements were easier to determine by simple mathematics. Forward moving speed was gauged by the time the tornado touched down from community to community, and the direct-path distance between them. Records were set for forward moving speed in the Tri-State Tornado, ranging at the highest from Griffin to Owensville, Ind., where it reached 73 miles per hour. Kern commented, "The average forward moving speed of a tornado is 25 to 35 miles per hour. The Tri-State Tornado then was moving almost three times the normal rate of speed of an average tornado."

By this, it might be inferred that damage could be lessened when linked to a faster moving storm—more forward speed, less time in the vicinity, less damage on the ground. Kern elaborated upon this school of thought:

"Nothing we know about that scenario would relate movement of the storm itself to the intensity. Slowing down or speeding up forward moving speed does not change the intensity of the winds within the vortex.

"What it does do is affects the intensity of the *damage*. If a storm moves through at 70 miles per hour, the actual time spent in a town or area is less. At slower forward moving speeds, it would stand to reason that the storm would spend more time in a town or area, creating more devastation by the fact that it was simply there longer."

This reasoning, however, did not stand up in Griffin. When the tornado was moving overhead, the storm system was traveling at a mind-boggling 73 miles per hour. Wind speed within the storm, then, would have had to be tremendous, having the potential to break the 318 mile per hour mark set by Professor Ted T. Fujita and Allen Pearson in the Fujita Scale. (This is defined as a scale used to rate the intensity of a tornado by examining the damage caused after it has passed over a man-made structure, and has since the creation of this book been adjusted, in 2007; see accompanying chart).

This is something that has been left up to the speculation of weathermen for decades.

"There would be no way we could possibly know the wind speeds of the Tri-State Tornado," said Changnon. "Only estimates."

No measuring device existed in those days, and relatively few now in existence would be sufficient to rate the top vortex speeds due simply to the fact that it would be difficult to locate positioning within the storm of the highest winds. It would be only by chance that precise measurements of the highest wind speeds would get taken. What would more likely be accurate would be to assess damage caused by the winds and set forth a range of wind speed that could have caused that particular damage. Types of houses and other structures, sizes of trees bent, twisted or broken, weights of large objects and the velocities at and distances to which they would have been hurled, were more apt to provide determination of wind speeds.

One remarkable aspect of the Tri-State Tornado remains the fact that it had an uninterrupted path of 219 miles from start to finish, and some recent research has turned up that upwards of an additional 15 miles, perhaps at a location *before* Ellington and Redford, may have actually been stricken early on, making it 234 miles (which theory will be examined in a following segment). Whatever the ultimate length, at the turn of the 21st century, it still held the record for the longest tornado track in weather record-keeping history. This phenomenon, while a unique part of the Tri-State Tornado, is not unique among tornadoes as a whole. Changnon has partnered with others and they have "studied twenty or thirty tornado tracks; there were four or five others, too, even in the state of Illinois, that had incredibly long ground tracks." The element that lead to the tornado staying on the ground for that extraordinary amount of time, noted both Kern and Changnon, was that of the trough of

Damage f scale	Little Damage	Minor Damage	Roof Gone	Walls Collapse	Blown Down	Blown Away	
	f0	f1	f2	f3	f4	f5	

Windspeed F scale	17 m/s 32	50	70	92	116	142	
	F0	F1	F2	F3	F4	F5	
	40 mph 73	113	158	207	261	319	

To convert f scale into F scale, add the appropriate number

		Little Damage	Minor Damage	Roof Gone	Walls Collapse	Blown Down	Blown Away
Weak Outbuilding	−3	f3	f4	f5	f5	f5	f5
Strong Outbuilding	−2	f2	f3	f4	f5	f5	f5
Weak Framehouse	−1	f1	f2	f3	f4	f5	f5
Strong Framehouse	0	F0	F1	F2	F3	F4	F5
Brick Structure	+1	–	f0	f1	f2	f3	f4
Concrete Building	+2	–	–	f0	f1	f2	f3

The Fujita tornado scale (F scale) pegged to damage-causing windspeeds. The extend of damage express by the damage scale (f scale) varies with both windspeed and the strength of structures. The bottom graphic shows the modern Enhanced Fujita Scale which is self-explanatory. Both scales were developed to assess wind speeds by the force it takes to either move the objects on the scale or create certain damage to structures.

ENHANCED FUJITA SCALE

EF0 (Gale)   65-85 mph | 3-second gusts

EF1 (Weak)   86-110 mph | 3-second gusts

EF2 (Strong)   111-135 mph | 3-second gusts

EF3 (Severe)   136-165 mph | 3-second gusts

EF4 (Devastating)   166-200 mph | 3-second gusts

EF5 (Incredible)   over 200 mph | 3-second gusts

low pressure driving the tornado forward. The low pressure created a "locomotive effect" in the atmosphere that day; winds in the jet stream were moving at fast, forward-moving speeds; winds within the supercell were moving at a high rate of speed as well. The two together, Changnon and Kern agreed, created a storm that was very stabilized and held together tightly by atmospheric conditions. Only after the storm began breaking apart, as it did outside Griffin, Ind., when three funnels became evident, did the destabilization begin to occur. By the time the Tri-State Tornado passed Princeton, Ind., it had lost the support of the high-speed jet stream winds and had begun to tear itself apart from bottom to top, the wind speeds in the cumulonimbus formation of the storm grinding away from the higher-rate upper level winds.

> *"We've seen events not dissimilar since the Tri-State Tornado. It's not particularly exotic in meteorology....we don't have to call upon some strange process to explain this storm"—Dr. Charles Doswell III, Doswell Scientific Consulting, Norman, Oklahoma*

Upon completion of the first draft of this book, there came to be another weather expert exploring the specifics of the Tri-State Tornado, and utilizing not only decades of experience, but advances in the field of meteorology to gain a better understanding of the event in light of 21st century weather analysis.

Dr. Charles A. Doswell III, originally from northern Illinois (Elmhurst/Villa Park area), graduate of the University of Oklahoma in Norman with a doctorate in meteorology, member of the American Meteorological Society and an incredible array of weather advisory, analysis and forecasting groups nationwide, distinguished speaker and professor, has been reviewing facts and figures of the Tri-State Tornado with a fresh eye and a view toward conditions on that March day in light of the latest understanding of cyclonic systems....and his understanding of it is in stark contrast to views previously- and long-accepted.

In 2002, Doswell joined a team of active and retired meteorologists and, in

Chuck Doswell

what he at times laughingly called a "self-funded field trip," began an exploration of the Tri-State Tornado's path, much like this book's author did: Knocking on doors, calling upon the kindness and services of libraries and historical societies to dig out decades-old original accounts of March 18, 1925, and taking out ads in newspapers serving the affected areas of the historical storm. He called the effort the Tri-State Reanalysis Project, and the dedicated group set out to learn as much as they could about the historic and devastating event.

Utilizing global positioning satellite (GPS) technology, the group (composed of team founder Dr. Robert A Maddox, Doswell, Charlie Crisp, John Hart, Robert H. Johns, Dr. Matthew S. Gilmore, Don W. Burgess and Steve Piltz) hit the ground in 2002. At what was the long-established, recorded starting point of the Tri-State Tornado, between Ellington and Redford in Reynolds County, Missouri, the team promptly found credible evidence that the first known damage from the Tri-State Tornado was at the

Shannon/Reynolds County line, about five miles, and possibly further, southwest, than was originally reported or believed.

"It appears that Shannon County was the starting point of the Tri-State Tornado," Doswell said. "We located a damage point right on the Shannon/Reynolds County line where there was a farm damaged. And there may have been an earlier, separate tornado from the same storm in Shannon County, although it appears the tracks didn't connect."

This may have been the reason why Jim Flowers, whose input was given in the very first chapter of the book, had been told by his father and other relatives for so many years that the storm, complete with funnel, appeared to have been building off in the distance southwest of the John Flowers residence, which was still several miles to the southwest of Redford. Nevertheless, experts over the years claimed that the tornado touched down between the Ellington/Redford area, as that's where Sam Flowers' body was found…and since he was the first casualty of the storm, it became a part of the authentic history to state that the storm *began* in essentially the same location.

However, Doswell said evidence shows that's just not the case; and that would mean that if the Tri-State Tornado began at the Shannon/Reynolds County line, the continuous *track* of the storm may not have altered much, but the *existence* of a storm at certain notable points "on the ground" adds up to another four or five miles to the 219…meaning 223-224 miles, and additionally, meaning that the history books may have to be adjusted to reflect this. And if the storm produced an earlier tornado—as some reports further into Shannon County, including in the area of a community called Gang, seem to indicate—there may even be more miles to add to the phenomenal length of time the storm was in existence.

Doswell also commented on other "lore" specifically regarding the meteorological aspects of the storm that his group uncovered (with Doswell and Burgess examining the Missouri track themselves; other members of the group spreading out across Illinois and Indiana).

"The 'lore' is that the tornado developed 100 miles west of the surface low that day; that it overtook the surface low and persisted at the center of the low for a significant portion of its life; and that it occurred at the front center of the parent thunderstorm," Doswell said. "Our analysis of the data negates those bits of lore."

Doswell said his group is continuing to develop factual data from data plot points they have been able to uncover, these obtained from obscure weather bureau station reporting regarding conditions on that day, but that it is a laborious process, and the data remained in the analysis process at the original date of this publication (2011).

However, Doswell did say that he has discovered, through this analysis, that the Tri-State Tornado did not form west of the synoptic low (the meteorological terminology for Changnon's assessment of the positioning of the developing cyclone) and it never co-located with the low center….meaning that the engine previously claimed as the driving force behind the power of the actual tornado was misunderstood for decades, even upon Changnon's assessment of it.

"The lore is that the tornado occurred on the front side of the parent thunderstorm," Doswell said. "The data shows that this isn't the case."

Doswell said he and his team have discovered to be fact something that many survivors quoted in this book described repeatedly: that the tornado, as part of the supercell, was a wedge-shape that at times did not appear at all to be recognizable as a tornado, commonly thought of as a snaking, whiplike, moving object.

"We've seen this at times," he said of the wedge-shape as well as of his more than four decades in the field. "The whole meteorological setting is in some sense familiar to us. We've seen events not dissimilar to this tornado. It's not particularly exotic in meteorology; we don't have to call upon some 'strange process' to explain it."

Where Doswell is in agreement as far as previous assessments of the storm is that of multiple vortices within a single vortex.

"With the continuity of the path, it's very possible for this to have been one tornado," he said, "but it's quite likely that it was multi-vortex. Virtually every tornado of that size and duration is multi-vortex.

"Now, there are some stories, like that of Biehle, where the surveyors claim it split and had two parallel tracks. We interviewed several people who said 'absolutely not,' that this didn't occur," Doswell said. He did not, however, discount Julius Hotop's report of how one twister snaked out from the main storm, danced around the Hotop farmstead, then slammed back into the storm as it receded away from Biehle.

"What Julius described is called a 'satellite tornado,'" Doswell said. "That happens in these large storms. But it did not produce a parallel track," he said, acknowledging the chaotic path the satellite twister took across and around the Hotop farm before rejoining the larger system. "I don't know how that 'parallel track' premise developed. It was written up in the original weather bureau report, but we have no evidence that it occurred."

Doswell also acknowledged satellite vortices in Griffin, where reports in the days and weeks following the storm from eyewitnesses at the time indicate that three vortices dropped over the western hills before heading into Griffin. Surveyors also found what appeared to be three tracks of vortices leaving Griffin, but eyewitness accounts from the time made the greatest impact on weather bureau reports.

Doswell didn't dispute the eyewitness reports, but did have comments to make about what folks said to those collecting information at the time of the days and weeks following the storm, as well as what he has experienced in the field, 're-tracing' the actual path that the system that spawned the Tri-State Tornado took.

"It's a real challenge," Doswell said, "because eyewitnesses speak in non-meteorological terms about a meteorological event….so we as investigators have to listen to what they say and place it within our realm. These eyewitnesses, they didn't know what they were seeing, so they came up with whatever references they could; they were correct in what they were seeing on March 18, 1925, but the interpretation of it, as weather experts, was probably wrong. So now the challenge becomes listening to those witnesses who are left, and interpreting what they say, and putting it in our terms."

One thing Doswell noted about the eyewitnesses who survived into the 21st century is that the event of the Tri-State Tornado "is indelibly burned into their brains.

"It's clear to me that this was a huge event in their lives and they feel frustrated because nobody would listen to them," he said. "So they can talk and talk and talk about it for hours, and remember more things about it the more they talked.

"We just can't know," he concluded, "unless we experienced it ourselves."

Doswell and his group are working toward producing a book about not only their experiences re-tracing the path of the Tri-State Tornado, but also specifically about the meteorological aspect of the storm. They are looking to produce new analysis in light of updated equipment and technology, and believe this will shed a whole new light on the most devastating tornado to ever strike the United States. Much like this book, they are doing this of their own funding, and research is a slow and painstaking process. They continue to look for any and all input those in the tornado's path, their children, grandchildren and great-grandchildren, may possess in these many years past the tornado's violent occurrence, and hope to make the experience interactive for everyone via a central data collection point, in hopes that the entire event may be reconstructed on a meteorological basis, so greater understanding of this and other storms can be had.

"It's a legacy we're going to leave with our profession," Doswell said.

*   *   *

Changnon, Kern and Doswell, along with countless contemporaries at the National Weather Service locations throughout America, have studied the storm and its devastating consequences for decades. However, with all of man's technology to discern changes and affects of meteorology, there is much still unknown about the formation of tornadoes.

Changnon was almost philosophical as he said, "When it occurred, there was so little solid knowledge of tornadoes. They were just huge, awesome ghosts."

Doswell surmised, "The jury is still out on the details of the structure and analysis of this event...but we do know that it remains one of the most important tornadoes in history."

Weather predictions many years after the Tri-State Tornado became more and more advanced and sometimes highly accurate in their assessment of the meteorological conditions preparing to affect a region of the country. Developments such as Doppler radar, which is radar 'lying on its back' and shooting waves upward in order to pick up disturbances in the atmosphere within its range were helpful, but not entirely precise. Even though Doppler is so precise that it can detect something as seemingly insignificant as a feather floating through a cloud formation, its reach was limited from the outset and funding to place Doppler in key areas was always a sore spot and of prime concern. In order to be completely effective, Doppler radar towers needed to work in tandem with others, and there were, at the turn of the twenty-first century, relatively few overlaps. Oddly, the very area that the Tri-State Tornado hit in southeastern Illinois and southwestern Indiana remained 'uncovered' by Doppler at the time of the interview period for this book (1999-2000).

What remains to be done with regard to the safety of those who are in the path of such violent weather is generally up to the individual, for only so much responsibility can befall those who predict and warn about the weather, and those who fund those prediction and warning operations. Perhaps many still act foolishly when dangerous weather threatens, by going out into the elements and standing and watching until the menace is upon them. But most storms, said both Kern and Changnon, kill by surprise. The hazards that the Tri-State Tornado created weren't all visited upon those who were outdoors watching the massive storm approaching. Many were going about their business until the walls fell in upon them, or the fires took hold. Many were making the effort to take cover and did not succeed, unaware that what was approaching was out of the normal realm of storms and would not just blow past them to be watched in its retreat.

In the United States, more people are killed by lightning strikes than by tornadoes, which number into 1,000, on an annual basis. But some day, if weather conditions are just right (or wrong), the likelihood of another storm the size and magnitude of the Tri-State Tornado could very well occur again. It is the hope of this writer that the nation as a whole is prepared for such an eventuality, and that the awareness of such a possibility is kept at a high level when each storm season approaches and runs its course.

Chapter Nineteen

The Sandborn Story: The Tail of the Wind

At about the time interviews for this book were being completed, and right after the 75th anniversary of the Tri-State Tornado, I received a phone call from a man who had read about the tale of the tornado in a press release I'd sent to all the local newspapers in an effort to locate more survivors, particularly those in the area affected by the storm many decades before.

The man, a Rev. Melvin Rose, left a lengthy message on my machine. He explained to me that his mother, while not necessarily a survivor of the storm of 1925 as she was 35 miles northeast of the path it took through Gibson County, was reported to have known a survivor; or at least, someone who was a potential survivor.

At fourteen years of age in 1925, Goldie Rose was Goldie Newkirk, a lifelong resident of Sandborn, Indiana. Sandborn lies to the north and east of Princeton by more than 35 miles on crookedly running highways, which, in their ambling, provide scenic routes along the drive. Over the years, Goldie had related a tale to her son, Melvin, about what happened to her and those she knew on that fateful day of March 18, 1925, when Death took to the skies and rode across three states in the form of killer winds and cyclonic terror.

Melvin, a man of God, insisted he had been raised a decent Christian fellow, in fact, all of Goldie's children had been taught the fear of God and of what would possibly be in store for them should they leave His good graces. One of the things Goldie instilled within her children was the fact that they must never, ever lie. The Bible told in very clear language that what awaits liars, as well as fornicators, adulterers, those who speak falsely against the Lord and a host of other well-explained transgressions of sinners, was the lake of eternal burning and cutting-off, a place of fire. It existed not as a place of unbalanced punishment, meant to burn and burn and torment for an eternity those who'd experienced only a lifetime of sin. Instead, it would consume everything and cause it to be as though it had never existed at all, so complete was its destruction that it literally wiped you out of the mind of God. And if you were gone in the mind of God, you were gone forever.

This was what Goldie taught her children.

So they grew up never to lie, and she grew old never to confuse them with a lie. Her word, like her name, was Gold, and her children and family and others who knew her knew well enough that what she said, if not the gospel truth, was at least the truth.

And what she'd said to her son Melvin over the years was what was related to me that day in late March 2000 when the final interviews of this book were being prepared to be written in chapter form and the stories, I thought, for the most part, had been told.

The story remains allegory, a local legend, oral history that after more than seven decades had faded from the memories of even those in Sandborn. In the year 2000, Goldie, removed from her home in Sandborn to the Linton Nursing Home in Linton, Ind., north of the town by mere minutes, was no longer as conversant as she had been mere months earlier, I was told. When asked about the storm of 1925, she brightened with sparks of recognition at the date, but those sparks did not produce flames of memory as they once had. She recalled, however, that her father and some other men of the village were involved. And she recalled, yes, they at the school she attended were made aware, that evening and more clearly the next day, that something terrifying and beyond comprehension of young, school-age children who knew storms as only thunder-boomers that rattled the walls, had happened to the towns and villages to the south and west of them, and they in Sandborn were the fortunate ones.

Beyond that, her recollection faded into the years of her age and other, more vague and immaterial incidents of bad storms that swept that part of the state of Indiana. And the Big One of 1925 took a back seat to others, seeming to elude so many of the age of remembering when it came to the turn of the twenty-first century.

In the Sandborn area, there were many around who had been small children during the time of the storm in question. However, none seemed to recall with any accuracy the whys and wherefores of the nightmare twister which had decimated a small area of their state just to the south and west of them. Further, none remembered the story that Goldie Newkirk Rose had been telling for years. But so many of these years had passed that this was not surprising. It became clear to me that if I were going to tell Goldie's story, I would be telling it as a story and not as an account; I would have to dress it up, as it were, take a few literary liberties with it. The sole person from Sandborn who would deliver the information would be Goldie herself, what little she had to tell; and in the limited way she was able to tell it, there honestly wasn't much to go by; so, much of it would have to be my imagination, hence the telling of the tale having Goldie run back and forth in the account (in actuality, I later learned, Goldie had suffered polio as a child and by 1925 was largely confined, unaable to carry out such typical childhood antics, sadly enough). It would have to be said, therefore, that it was merely local legend, recalled by a woman whose memory of it was passing with each day.

Her son, however, remembered what he had been told through the years, when Goldie's faculties of memory were not vague at all and her mental acuity was sharp. He and those around him had no reason to doubt that what they were being told was the fact of the matter, not local legend, oral history, allegory—folklore.

And this was how it was related to me.

Sandborn, Indiana
March 18, 1925 – appx. 5:45 p.m.

Goldie Newkirk was headed in from tending to the cattle in the barn her father had recently shorn up. The shoring up had had to be done; poles supporting the hayloft had begun to bend last year with a swayback look like a riding horse gets at about nine years of age, and Newkirk was no longer comfortable with the very roof over their heads at any time any of them, especially his five children, stepped foot in that barn. So last year, the ceilings were jacked up in the areas around the poles, and new, hardwood timbers were brought in, complete with a few trusses crisscrossing the ceiling for added support. It was good, Goldie Newkirk thought, that her father had done this. The weather that season was showing itself to be worthwhile and cooperative, regarding warmth and consistent rainfall. The loft was soon going to be full of sweet smelling, animal-feeding hay. Goldie herself couldn't wait to get up in there and play in it, jumping and rolling and sliding most of the summer and well into the fall.

The sun was almost set. There was a faint glow in the western horizon straight ahead of Goldie as she left the vicinity of the barn and the barnyard. About an hour ago, the sky had darkened to the south with a dusky blue shade that had crept along the horizon until it was completely gone and things were clear again. Goldie didn't mind storms. But she was glad that this one, however threatening it may or may not have appeared, had moved on its way and left the skies around Sandborn clear once again.

She'd felt a slight breeze at that point, and had stood outside the barn stalls with the milk cows gazing at her dumbly as she'd peered intently to the south. The black clouds had barely broken the horizon. Goldie had felt them more than she had seen them. There was something wicked in the way the clouds had crept stealthily along the skyline, just above the surface of the ground almost, as if they'd been trying to keep out of Goldie's sight. But she'd caught them out of the corner of her eye the minute they'd appeared in their sneaky shifting crawl, and Goldie had wondered what the people under those clouds, probably all the way down north of Petersburg, were seeing at the time. She had gone into the barn once the clouds had disappeared, and had for the moment forgotten about them. It was only when she emerged into the pending sunset that the vision had been recalled, and Goldie shuddered briefly over it as she walked the well-worn path back up to the house.

She had just latched the gate behind her when she heard a terrific *whoop!* come from south of the farm.

Her heart instantly jumped up into her throat.

It was her mother's voice making that sharp, unsettling cry, and she had heard it only once before, but the memory was etched into her brain like a branding iron on a bull's flank. It in fact had been a bull that had prompted that war whoop one time before out of her mother, when, having gone out into the pastureland that bordered the immediate farmstead, her mother had been secretly followed by the next-to-the-youngest child, four at the time and lacking the sense God gave a goose. The little one had slipped under the fence, and had gotten into the holding pen for the stud bull, which came complete with curved, sharpened horns. The bull had noticed the child before Goldie's mother had. And the speed at which Goldie's mother had been over the fence, retrieved the child, and exited the holding pen once again had been almost as quick as her sudden *whoop!*, if her account was to be believed. And the Newkirks had no reason to doubt their matriarch; she was as stoic and solid a Christian woman and devoted mother as one could ever find.

But here was that cry again. Goldie hadn't gone running the first time she'd heard it, but there was nothing to prevent her now, since she knew her mother would never make such a sound without good reason. She picked up her skirts, which she hadn't changed out of after school, and began a stumbling run to see what the matter was.

Off in the distance, near what used to be a duck pond and now was merely a watering hole most of the year for the cattle and horses and remaining quite stagnant as a result, were her father and mother. The elder Newkirk, Francis, was on his knees leaning over the meager bank of the former duck pond. The family matriarch, Mae, was running in little desperate, frantic circles, occasionally tossing her hands in the air in a manner that could only denote hopelessness and despair.

"Ma, what *is* it??!!" Goldie cried in a voice so high-pitched and frenzied that she scared herself with it as she ran. "What?? What??"

Mae Newkirk suddenly took notice of her daughter and stopped cold in her tracks.

She began to run across the hardpack pastureland that surrounded the watering hole, but recent rains made the ground uneven and mushy and Mae stumbled much as Goldie had upon her initial sprint. Her intent, it was clear, was to stop Goldie from progressing any closer to the duck pond, but Goldie had already ascertained her mother's purpose. Nimbly, she dodged the woman and made for her father, who was still kneeling in the muck and mud of the nasty watering hole.

Newkirk was up to his elbows in mire and was pulling on something equally buried in the goo, covered and coated with the slimy substance. For a moment Goldie imagined it was a calf, having slipped on the little bank and tumbled into the mere depth of about three feet—but enough to suffocate it were the mud to be as thick as it tended to be in a mild spring. Then, with a spark of clarity, Goldie realized there were no calves, they were all due to be born within the next month, but none had shown themselves yet. Puzzled, Goldie almost tiptoed toward the figure of her father on the bank, wondering absently if a neighbor's calf may have been already born and wandered upon the Newkirk's property, or perhaps even a foal she had heard was born up on a farm to the north.

But her mother's insistent sudden grasp upon her arm was telling her otherwise. The normally unruffled Mae whipped Goldie around to her and clasped her to her bosom, staring down into the child's face with an intensity Goldie could not know was reserved for survivors of war and other hideous events a gentlewoman should not see in her lifetime.

"Don't go over there, Goldie," Mae Newkirk hissed. *"Don't."*

There was nothing exclamatory about Mae's statement. It was a fact, a declaration. There was no way Mae Newkirk was going to let her fourteen-year-old daughter go near that duck pond.

And yet Goldie had seen her father floundering, struggling with something in the pond that needed to be out more than it needed to be in, and Mae was doing nothing to promote that eventuality. Goldie glared at her mother in a way that she had never before in her life dared. In her eyes she said it all. She was going to the pond, and she was going to help her father, no matter what the cost between her and her mother. Whatever was in that pond was going to be out, and with Goldie Newkirk's help.

She wrenched away from her mother and quickly moved out of her grasp so that she would no longer be able to detain Goldie further. Mae Newkirk gave up and tossed those hands once more, helplessly, into the air, and with another shout, ran toward the farmhouse and a handful of neighbors who had heard the commotion and were now running to sate their own curiosity.

Goldie did not wait for them to reach her and her father and whatever it was that he was struggling with in the pond. She hurried to him, slowing only upon seeing what he was gripping, almost with futility, in his muddied hands.

There was a man in the duck pond.

He was caked and covered with mud, and it glowed with a slightly greenish tint in the reduced sunset light. As Goldie approached with an almost overbearing trepidation, she noticed little pieces of twigs and leaves and other debris embedded in his hair, maybe, she realized with a lurch of her stomach, even in his skin. He was completely unclothed. Her father by now had him gripped firmly about the waist, and he was attempting to clear the man's nose of muck and slime, but the task seemed impossible for just two hands.

Newkirk in his struggling turned to his left and saw Goldie inching forward, frightened but drawn to the sight.

"Help me, Goldie!" he gasped. "I think he's still alive!"

Newkirk staggered back with the weight of the man, who looked to Goldie to be somewhat smaller than her father, who stood at least six feet tall and maybe weighing in at close to 180 pounds.

"It don't matter about your dress, now," he managed to say, "we gotta get him out of here. He's probably bad hurt. Looks like he got beat up and dumped in here, and it's our property, so it's our responsibility."

Newkirk needed to say no more. Goldie ran to her father, the feeling of accountability for what was happening upon their own land outweighing any fear that may have been present. She reached for the slimy arms of the man and helped her father haul him away from the bank and onto more stable ground. She wiped her hands on her skirt and pinched the man's nose to force out the grime and muck from the pond's bottom.

"Turn him over," Newkirk whispered as if he were listening for breath or some utterance of sound from the filth and the man inside.

Goldie helped her father shift the man to his side, and as he lay with his left arm beneath him, Goldie found it necessary to prop him up to keep him from flopping either over on his back or face-down in the dirt. And as she did so, she detected the slightest bit of movement from the area of his ribcage where she held him with her hand gripping under his armpit.

"Daddy—!" Goldie whispered. "He's trying to breathe!"

Newkirk let go of his hold on the man's head and neck and positioned himself near the mud-covered face. His ear rested fractions of an inch from the nostrils of the man, which leaked slime, snot… and yes, now, air.

Newkirk's neighbors arrived at that moment.

Goldie arose when Tom Pedersen, a kind, elderly man from the adjoining farm to the east of theirs, knelt beside the three of them. She'd begun to feel faint, seeing that muck and mucous drizzling from the nose of the stranger.

"What happened?" Pedersen asked in a gruffness that was intended to belie the concern he obviously was feeling.

The confusion in his normally twinkling blue eyes, those that Goldie and her entire family had come to be as familiar with as a grandfather's (for indeed that was what Pedersen was, many times over), was clouding his otherwise pleasant demeanor.

"Don't know," Newkirk answered, equally confused. "I was out there along the fencerow checking where the gully had washed a post aside, and I heard a sound, and when I turned around…"

Newkirk heaved a great gasp of breath, as if he were the one who had inhaled the mud instead of the man who now lay gasping on the ground.

"What kind of sound did you hear?" Willie Cain, another area farmer but one much younger than either Newkirk or Pedersen, felt compelled to ask.

Newkirk looked up from the man, who was now moving his head from side to side and moaning softly.

"It was a splash, Willie," he told the young farmer. "I just heard a big splash, and when I came over here, there he was, about halfway on the bank."

Newkirk shook his head, and Goldie, who had stepped back several inches to allow the men access to the mud-man, now moved up again only to hear her father's words and become overtaken with a chill at them.

"It was like he fell out of the sky or something," Newkirk offered.

The two neighbor farmers glanced at each other, and both knelt down next to Newkirk and his muddy guest.

"Let's at least get him up to the pump in the barnyard where we can clean him up," Pedersen said. "Goldie, you go on ahead to the house and bring back some of your pop's overalls and some of Ma's old sheets or towels, okay, honey?"

"Okay," Goldie whispered, already breathless with the thought of racing the long trail back to the barnyard and house.

She turned to go, and then suddenly, a thought flickered across her mind that deactivated her feet and legs long enough for her to ask.

"Shouldn't I have Ma go get the doctor in town?" Goldie suggested, feeling immensely proud and adult-like that she herself had thought of the notion.

But those fine feelings abruptly faded when she got a good look at the expression on Tom Pedersen's weathered face.

"Won't do any good," Pedersen said with mild uneasiness creeping into his voice as he looked down and the mud-covered man and his savior, who was carefully dabbing the muck away from the muddied nose with his kerchief in order to promote life and breath.

"Why?" Goldie asked, certain she was not going to like the answer.

"The doctor's been called down to Princeton about an hour ago," Pedersen said carefully. "Doctor from up in Linton, too."

Once again, Goldie asked, although she knew now that she didn't really want the truth.

"Why?"

"Bad storm down there in Princeton and Owensville," Pedersen responded. "Took out the Southern freight yards and the factory in town. And they said there ain't a building standing in that little town by Webb's Ferry called Griffin."

Goldie Newkirk studied the elderly man's face for a moment, but she wasn't seeing it. She was instead seeing the vision of that blackened line of clouds on the south horizon, having been there only moments before, crawling as if on their belly along the tree line as they soundlessly edged their way out of her view. And she wondered, if only briefly, what murderous secrets they had carried with them as they traveled to their unknown destination.

She turned and ran, tears streaming from her eyes, toward the farmhouse.

March 19, 1925 – 6:30 a.m.

Daylight was breaking when the doctor rode in from Princeton in his Model T Ford, a 1925 car that handled the rough dirt track toward the Newkirk farm with all the ease of a year one jalopy.

The doctor was rattled enough, he told the farm owner Francis Newkirk. Doctor Talbot Smith, who was a Great War veteran and had seen massive combat injuries, was a pale and shaken man when he arrived, having been summoned in the predawn hours by neighbors in Sandborn who'd made the trip into Princeton to find him. Dr. Smith had seen more piercings, more shattered bones, even a couple more beheadings in Princeton than he'd seen in all his months of duty on the battlefields of Europe. Bloodied himself from all the makeshift surgeries he'd had to perform, Dr. Smith had collapsed on a cot in the sanatorium on State Street until someone had awakened him from slumber, telling him he could go home, but on one condition:

That he examine the man who'd mysteriously appeared in Francis Newkirk's duck pond the night before.

Another Sandborn resident, Lamar Ikes, had come with his brother to fetch the doctor and volunteered to drive him back in his own car. Lamar's brother Edgar took the Ikes truck back to Sandborn, followed by Lamar driving the doctor's Model T. This was the small caravan that made its way out of severely damaged Princeton and on northeast to Sandborn, to arrive at about 6:30 a.m.

Dr. Smith had experienced only four hours' sleep, including the hour and a half it took for Lamar Ikes to drive him home. He was bleary-eyed when the Newkirks met him on the front step of their home. But the bleariness dissipated quickly with Mae's Newkirk's good strong coffee and an extended examination of the patient, who by now was awake and proclaiming hunger and a need for coffee himself.

Goldie Newkirk hovered near the downstairs bedroom that was built as an afterthought to the rest of the house when it became apparent that Mae Newkirk was going to have Goldie's little brother, Ray. As a result, the doorway to the room still had with it that new-lumber smell; Goldie didn't necessarily know what the wood was, maybe pine, maybe oak, but it was hand-hewn and rough, even after four years, and smelled delicious in the early morning hours when there was still a chill in the air.

Goldie walked softly on bare feet from the foot of the stairway to the entryway of the hall that lead to the kitchen, and tilted her head in that direction. She could hear her parents talking, but not what their conversation was specifically about. She didn't need to. The topic was no doubt the man in the room behind her, being questioned by the doctor who fourteen years earlier had helped bring her into the world as Francis and Mae Newkirk's first child of five.

Goldie sensed movement from the kitchen and, not wanting to be caught, she darted soundlessly into the little anteroom behind the cooking area that had a door leading to it from both kitchen and hallway so it could be used for maximum storage from both areas. The room was shaped like a pie-wedge, being the corner of the house. It was closed-off with no windows or ventilation and doors that remained, for the most part, shut on both sides. But this morning the hallway door had been open because Mae Newkirk had been washing linens and towels most of the night last night, cleansing them of the mud from the man whose name was not yet known.

All the linens and towels in the hallway side of the pantry were gone, and the door propped open, as they were drying in the sunlight of dawn, having been hung overnight by Mae and Tom Pedersen's wife, Iris, who had come to help. The mud had been caked onto him hopelessly, and doubtless still lay in cracks and crevices on the man that it shamed Goldie to imagine at her tender age, body parts she knew her father had had to wash because her mother simply wouldn't touch the privates of that man. But it made her uncomfortable at the same time, for two valid reasons: one, she herself had loved mud when a toddler and had foraged in it for mud pies, mud puddings and other mud concoctions that made her mother almost stark raving mad with the mess, and two, even in those times up until the age of about nine or so, utterly caked with mud herself, having deliberately ground it into her skin, she hadn't gotten as filthy as that man had turned out to be. It was almost as if the mud had been *driven* into his skin by some unimaginable force that had left him for dead in the duck pond.

Goldie's parents, eyes bloodshot and skin pallid from the sleepless night they also had spent, came softly down the hallway next to the staircase, and were headed straight for the bedroom where the patient sat with his doctor. Goldie pressed herself against the now-empty shelves in the pantry and closed her eyes, knowing she didn't have enough time to shut the door in front of her to keep her parents from spotting her tiny frame. She hoped they were too preoccupied with their unexpected guest to glance to their right and notice her, and with that hope she noticed she was breathing rapidly, shallowly. She forced herself to calm down and take deep breaths, inhaling the rich cinnamon-apple smell the pantry always had in it, for on the kitchen side, there were cooling racks for pies and most recently mother had made apple. Absently, Goldie hoped next week would be custard, she liked that kind so much…

Outside, she heard the voices of the doctor and her father, and they were notching up in volume as the exchange between them grew.

They were standing outside the bedroom door; Goldie didn't know why. She thought she'd smelled coffee as her mother walked past, and likely the brew the mud-man had requested was being delivered; she was certain she could hear, even through the door, the clinking of a coffee mug with a pot being tilted and poured against it.

"I'm not staying," Doctor Smith was telling her father. "The man's obviously either insane or has been beaten so badly that he's lost his mind from it."

"I don't understand," her father responded. "What is it he said that's got you so upset? Did he tell you who beat him? Is it someone we know? Someone who wouldn't ordinarily do this?"

"Frank, it's nothing like that. I've just had enough. This damned tornado that went through Princeton yesterday, and the hell I saw while I was there; I guess it's taken its toll on me. I'm not a young man anymore. I can't sort out what's real from what's not.

"All I know is, I saw what was real last night in that hospital, and that was just too real. People with two-by-fours through their gut, up walking around, refusing to lie down and die decently. People with arms and legs that looked like they'd been twisted off, some of it just clean amputation of the bone with just a little twist of skin at the end, and feeling no pain in it. And bodies, parts of bodies, and bodies with no heads. Why, someone told me there was a woman left sitting in her chair in a house in Baldwin Heights with her knitting still in her hands, and her head was gone at the neck, a piece of glass from the front window took it clean off."

The doctor took a shuddering breath and Goldie, whose stomach was verging on shifting and lurching following the doctor's comments, could hear nothing for a moment, not even the clink of the coffee mug, and then he said, "People shouldn't tell such tales like this man is telling. He has no right."

Goldie heard what sounded like the doctor gathering up his bag, and he took steps toward the front door. But he paused, and, just before she heard the door swing on its hinges, Goldie heard him say one more thing to her father.

"And if he's telling the truth, Frank, then God help him, because I don't know any way to."

The front door shut, and shortly, Goldie heard an engine start up and the Model T Ford headed back down the driveway from her home.

Her father's footsteps approached the main hallway again, and began to recede toward the bedroom. But just as she was certain she could escape the closet out the other side to the kitchen, Goldie saw a rugged hand push the door to the hallway open a little wider, and her father appeared in the pale light of dawn streaming from the window at the top of the staircase to her left.

She stared at him with huge eyes, looking for all the world like a trapped rabbit.

"Come on, Goldie Marie," Francis Newkirk told her in a gentle voice. "You helped me reel him in, the least I could do for you is tame that curiosity a bit."

He took his daughter's hand and lead her to the room where her mother tended their catch with strong, almost acrid coffee and was asking him in a voice much calmer than yesterday's if he would like some biscuits for breakfast.

"Howdy," Newkirk announced as he entered the room. The mud-man acknowledged the greeting by a nod, and without a trace of a smile. "I'm Frank Newkirk, and this here's my daughter, Goldie. We was the ones who got you out of that mud-hole yesterday."

341

The mud-man extended his left hand, which did not grip the ring of the coffee mug as did his right, and offered it to first Newkirk and then his daughter, without so much as a word.

"Do ya remember being in it?" Newkirk asked of him, and again, the mud-man sat upright in the small frame bed, intended for the little boy, Ray, who was asleep in his parents' room. The stranger greeted the question with a stony silence.

That was, at first.

He looked down into the black of his coffee, steaming in the mug, and seemed to study the way the early morning sun shot light across the surface of it, as if he were trying to figure out how this neat trick were done with the play of light on black.

"I do remember some," the mud-man said slowly, and Goldie looked at her mother, who was staring at the man, then to her father, who was doing likewise but almost scowling.

As the man went on, Goldie began to understand why.

"I do remember being able to breathe, and thanking the Lord for it. In my heart, of course; I had too much stuff in my mouth."

Goldie puzzled over the man's words. When he said "being," it came out "bean;" "breathe" was "breed;" "thanking" was "tanking." Goldie tilted her head to one side and narrowed her eyes. It was an accent. She'd heard such a thing when they'd gone over to Haubstadt last year during the festival. They were cooking bratwurst and sauerkraut, the worst smelling way to cook cabbage she could think of. It was a big party and there were even rumors of beer being made and sold there somewhere. It had been a German festival, and this man was speaking with a slight German accent, just enough to notice, and Goldie thought, if he weren't so tired and beat up, he'd probably be able cover it better than what she was hearing now. Her father had caught it, too. He was peering at the mud-man suspiciously. Germans were okay to some of the folks in southern Indiana, but a lot of people just didn't like them, thanks to the Great War, and many were hard-pressed to keep their German heritage, like their accents, from being noticeable.

"You're not from around here, are you?" Frank Newkirk asked slowly.

"Where's here?" the man returned just as slowly.

Francis Newkirk rubbed the palms of his hands on his overall legs, and went to sit down in a cane chair next to the bed, where Doctor Smith had been sitting minutes before.

"Why don't you know?" Newkirk asked, the distrustful edge of his voice unnerving to Goldie.

The mud-man was bewildered, however, and ignored the suspicion in Newkirk's words.

"Just don't," he replied, although it came out "Jus' doan."

"What's your name?" Newkirk asked.

The man pondered the question heavily for a moment.

"I don't know," he answered, and began to sound concerned, his voice rising in pitch on the last two syllables.

"You don't know who you are?" Goldie's father questioned.

The man seemed to be startled by this revelation, and he shook his head gravely, his eyes wide.

"No, sir," he answered as politely as his obviously rattled nerves would allow him.

Francis Newkirk leaned forward in his chair and Goldie saw that his jaw was set firmly, but it wasn't, she began to understand, in anger. Rather, it was grave concern that something was truly amiss with this situation, and they were all about to find out why Dr. Talbot Smith was so anxious to leave the Newkirk house and their uninvited guest behind.

"What *do* you remember?" Newkirk asked.

The man gazed out the window at a fabled blue sky as morning dawned bright and beautiful, and he began to speak

"I do 'member plowing the field." He took a sip of coffee, then a swallow, then a gulp. "The mules was acting up. There were two; I usually don't use both, but it was such a rocky stretch of groun', you know. I needed 'em both."

He glanced at Newkirk, then Goldie, and went on.

"They was acting up because they knew. I didn't know, but they did. There was jus' too many hills behin' me for me to be able to see until it was on me, but them mules, they could feel it."

"Feel what?" Goldie dared to ask.

"A big storm," the mud-man said, and his eyes, now staring as if right through Goldie following her question, were chilling her to the bone. They were dry and blue, but old, as if he had just witnessed eternity and would never recover from it.

"Oh, the biggest storm I ever seen, little girl. Came over the big hills and down onto the field I was workin'.'"

Her father was shaking his head.

"Big hills," Frank Newkirk said in wonder. "There ain't no big hills around here."

It was true. The land around them was flat as a flapjack, perfect Hoosier farmland.

"Where's here?" the man asked tiredly.

Newkirk tilted his head for a moment, then apparently decided to play along.

"Sandborn," he answered matter-of-factly; it was, after all, his hometown.

"Where's Sandborn?" the man asked, and for a moment, it appeared as though Mae Newkirk was going to burst into laughter at the ludicrousness of it. But she could see he was serious.

"Is it near Old Appleton?" he asked further.

Newkirk shook his head, frowning.

"No," he said.

"Frohna?" the man asked of him.

"No," Newkirk answered again.

"Wittenberg?" the mud-man answered hopefully.

"No!" Newkirk answered with finality, about to have had enough.

The man glanced briefly at all three of them, and tears began to cloud his eyes.

"I've never heard of a Sandborn, Missouri, before," he said with great despairing in his voice.

Goldie watched her father's face turn white as the sheets that were hanging on the line outside the window; her mother's lye soap was strong, the mud had almost been washed entirely away, and those sheets were like ivory once again.

"This ain't Missouri, mister," Newkirk said, and he kept his voice calm, still, for the sake of the man and for the two women beside him who finally understood the horror that was dawning on the mud-man's now-clean face.

"It ain't?" he asked dejectedly.

"No, sir," Francis Newkirk told him quietly. "This here is Sandborn, Indiana. You're about two states away from home, friend."

Epilogue

At the turn of this century:

Ellington saw the tornado as something that happened so long ago there weren't many around who could even begin to tell the details of Sam Flowers.

Annapolis was stoic; their downtown was destroyed, but they had lost only four.

Biehle, with Ridge and the rural Missouri areas struck after it, was confused over what had happened; at 67 miles per hour forward moving speed, it had happened so quickly they'd not had time for a second look.

Gorham was shell-shocked, stunned.

Murphysboro was bitter.

DeSoto was crushed with the weight of the loss of its children.

Bush and Plumfield found themselves the objects of gratitude of Hurst, Royalton and Zeigler. At least the storm, the latter three reasoned, hadn't hit them.

West Frankfort survivors were mostly grateful because the twister had left only 20 percent of the town damaged, a relatively small amount by comparison.

Parrish was gone, obliterated, never to return.

Hamilton and White counties were awash in grief. Never, they noted, had such a weather phenomenon taken the farm community, and the weather-wise farmers, so unawares.

Griffin, Owensville, and Princeton felt as though the river had betrayed them, having come up and slapped many of the residents in the face with the mud and muck that usually meant their livelihood by way of the rich run-off soil. Instead of preserving them, it was hindering them from being recognized by members of their own families. It was infecting their wounds and keeping them from reaching a recovery on an immediate sense, in that things were going to get back to normal soon, things were going to be all right.

It was a lie.

For so many, even the possibly fictitious mud-man who "landed" in Sandborn, Indiana, thinking he was still in Missouri, things would never be all right again.

*   *   *

At the cemetery on the southwest corner of DeSoto, the three little markers at what appeared to be a common grave had revealed themselves, through the stories of Lillian Bridges and Betty Moroni, to be just that.

Those little girls were sisters of Betty Barnett Moroni, whose family lost not only their home and their livelihood on that day, but three precious things that could never be replaced. Martin and Minnie Barnett, parents of these little girls as well as two other girls and a boy who did survive the monster's attack, understood on an individual level what DeSoto was forced to recover from on a community level.

The community of DeSoto, as well as so many communities in the tornado's path, lost children, not just the individual parents. There were empty beds and cribs, dust gathering on birthday presents and unused little eating utensils, should any of them have been retrieved from the debris that the storm left behind of their homes, and cried out for the children who were laid to rest in the city's cemetery. When Christmas came around, it was a time for joy tinged with sadness. When Easter was marked on the calendar, and the thought of resurrection was to the forefront, it was noted as an awesome concept, but for some, resurrection was not so much a comfort as it was a reminder of a scar that, regardless of the passing of years, may never, ever heal.

The death of a child is marked throughout the year—on the death day, on the birthday, seeing another child at play, or walking past the rocker in which the baby was rocked. While so many schoolhouses were wracked by the ravaging winds of the Tri-State Tornado of 1925, and many communities lost many young ones, the citizens of DeSoto, it can fairly be said, knew this better, perhaps, than any other community. And by pulling together and being strong over the years, they knew, too, how to overcome the merciless darkness that threatened to swallow some on a day-to-day basis through anguish and hopelessness. The ability to overcome was learned through the fight with an unwelcome visitor from the skies, which succeeded in some places, but only threatened, in others, to wipe these communities from the map—
And failed.

About the Author

Angela Mason spent a quarter-century of her life in the music and performance fields as a vocalist, musician, actress and on-air radio personality while living in Illinois, Indiana, the Gulf Coast, Eastern Seaboard states and even Germany. But the love of writing and of history overtook that and she has been published with or working for various newspapers, magazines and journals since 1992.

Death Rides the Sky is her first full novel-length, non-fiction book, and the creation of it helped her get over her fear of tornadoes, which was developed over several years of intense storms sweeping across the woods and fields of Wayne

The author Angela Mason with Tri-State Tornado survivor Mary Belle Melvin.

County, Illinois (where there haven't yet been enough earthquakes to warrant a book and subsequent facing of *that* fear…yet).

She resides in her beloved Southern Illinois with her husband Jack and a passel of spoiled cats, is mom to three beautiful adult kids and grandma to (so far) three of the most amazing granddaughters on earth.

Index